To Bill & Family

Badger Behaviour
Conservation & Rehabilitation

70 Years of Getting to Know Badgers

George E. Pearce

Pelagic Publishing
www.pelagicpublishing.com

Published by **Pelagic Publishing**
www.pelagicpublishing.com
PO Box 725, Exeter, EX1 9QU

Badger Behaviour, Conservation and Rehabilitation
70 Years of Getting to Know Badgers

ISBN 978-1-907807-03-9

Printed in the UK by the MPG Books Group, Bodmin and King's Lynn.

British Library Cataloguing in Publication Data
A catalogue record for this book is available from the British Library.

This book is dedicated to my late parents
for teaching me how to understand animal behaviour

Acknowledgements

I wish to express my gratitude to Mike Hughes who inspired me to write this book. Without his hard work and journalistic expertise, this book would not have been written.

To my daughter Pamela, whose dedication and encouragement has been invaluable. This book would not have been finished without her.

To my friends Marion Kuipers, Anne Smith, Margaret Bray, Francis Haynes and my sister Nancy Jones for reading the manuscript at various stages and urging me to get it published. And to my dear friend the late Chris Thomas, whom I was very fortunate to work alongside for a number of years. I shall certainly miss our stimulating discussions.

To my son Tristam, whose sharp eyes and quick reflexes have been a tremendous aid.

And, finally, to my loving wife Cris, for her patience, understanding and unwavering support.

Contents

Introduction

I was about seven years old when I first began to learn about wildlife, foxes especially. When the orchard where mother reared her young chicks, ducks and goslings was raided by the fox, she would call, "George, get the spade", and off I'd go to bury the chicks the fox had killed. I remember the spade was as tall as I was! I remember, too, looking for clues along the hedgerow and seeing a track through the briars and red hairs snagged on them. That was the trigger, the start of my interest in natural history, and then, as now, I wanted to learn more.

Growing up, I had heard folk talk about badgers and foxes, but not so much about badgers as today; then, it was mostly foxes. I believed the things that my father and fellow farmers said about foxes and badgers at that time, but I was soon to discover that most of it was myth. As a farmer's son, I had the run of all the fields in the neighbourhood and by day, as we carried out our work in the fields, I began secretly to study badger and fox tracks and the various signs that they leave behind. At night, when all my work was done, I would slip away quietly and sit or stand by a gate or under a tree and, piece-by-piece, I was able to uncover the secret world of the badger and the fox. I had no interest in reading as I found it difficult and throughout my school days I had problems with both reading and spelling. I now realise I am dyslexic, but this condition was not recognised in my school days of the 1930s and 1940s. There was no television then and in those days little time if any was allocated on radio to wildlife matters. So, it was a matter of watching, observing and learning at first hand from the animals themselves.

That was more than 70 years ago. Today, I am still learning and am still fascinated by badgers and foxes – both highly intelligent animals. My first encounter with a fox was one evening in August 1944. I was standing by an oak tree when I spotted a fox walking very slowly along a hedgerow, stopping from time to time, reaching into the hedge to bite off a blackberry. Not a single leaf moved, so gentle was it when picking the fruit. Fellow farmers had said nothing about foxes eating fruit. Perhaps they did not know then what we know now, that fruit in season accounts for a high percentage of the diet of both badger and fox.

I saw my first badger when I was cycling home one bitterly cold moonlit night during the 1946–7 big freeze; I was 14, and three miles from home. This first brief encounter was enough to kick-start what for me has become a lifetime's fascination. I did not know it then, but I was only yards away from an active sett; moreover, a sett that was on land I was eventually to own.

Although it has jaws powerful enough to crush bones and sever tree roots, the badger's preferred food is the soft-bellied, almost gelatinous, earthworm. Badgers live in small family groups, yet where conditions allow, they often dig extremely large, highly complex setts. The badger is a sociable creature with a laid-back lifestyle, yet like all creatures, it does what it has to do to survive.

Most people, even those who live in the country, see live badgers only rarely. Many see them only as corpses by the roadside. Thousands die that way every year, killed as they follow age-old paths that were there long before our roads filled up with speeding traffic. Others, I am sure, die as they search the verges and the roads for worms, insects and carrion. Forty years ago, when on the road, the badger played Russian roulette with one bullet in the gun. Now, with many more vehicles on our roads, travelling at very high speeds, it is playing with the equivalent of five bullets in the gun. The badger's natural predators may have long since vanished, but make no mistake, the car has a huge effect on badger populations, just as it does with other forms of wildlife.

I count myself privileged that, for most of my life, I have had the luck and the opportunity to watch badgers at close quarters, alive and well in the wild. I have also shared my home with them, caring for orphaned cubs, and helping to restore injured adults to health, some of them road casualties, others the victims of snares and gun shot. My days as a rehabilitator ended some years ago as RSPCA animal hospitals increased in number. But, I learned a lot from that experience and I am thankful that, in return, I was able to release back to the wild more than 100 badgers and over 60 foxes, as well as weasels, stoats, polecats, owls and kestrels.

Today, I earn my living working as a badger consultant, advising others faced with problems that centre in some way on badgers. Later in this book I shall explain in more detail what it is that I do and what can be done when badgers get in the way of what we call progress. As you will see, it is work that has not only given me an unusual insight into the secret underground world of the badger, but also first-hand experience of how it lives alongside and copes with, dramatic man-made changes to its environment.

There are many who question the role of the badger consultant. But, as I shall try to demonstrate, it is work which, in my case at least, is at all times driven by the need to give the badger a fair deal and to ensure its long-term survival.

Chapter One

A lifetime's experience

Allow me first to set the scene. How did George E. Pearce, farmer's boy turned pig breeder, become a badger consultant? It was definitely not a role I had planned, dreamed about or ever thought I would fill. In effect, I grew into it, with fate (in the guise of plummeting pig prices and rising animal feed costs) giving me rather more than a gentle nudge.

I do not have a college degree or any written qualifications – my classroom was the countryside, and my teachers were the animals themselves. Over 70 years, I have watched and watched, and watched some more, and my studies were the thousands of hours I put into observing badgers in all sorts of locations and in every kind of weather, from the hottest July evening to –9 °C in January. I have lost count of the times I have sat through the night at a sett, the times I have tracked badgers as they left their setts to forage, and the early mornings when I have waited for dawn to break to watch them troop back. In all those years, I have seen badgers grow from cubs to adults, seen setts grow, decline, fall into a state of decay and then come alive again. I have spent countless hours acquiring field-craft and tracking skills: I have dissected badger droppings; carried out autopsies; rescued snare victims; worked with the police and the Royal Society for the Prevention of Cruelty to Animals (RSPCA) to bring baiters and diggers to justice; and worked alongside my wife Cris and my children to rehabilitate injured and orphaned badgers. I helped to launch the Shropshire Badger Group and in all sorts of ways, put my long years of stockmanship to good use in learning more about and helping this rarely seen, and often misunderstood, creature.

Farming was my work, badger and fox watching was my hobby, but suddenly it all changed. Pig farmers (unlike those in grain, beef, sheep and dairy cattle) have never had the support of significant subsidies from the European Economic Community (EEC) and when feed prices shot up and low-cost imports increased, pig farming of the kind that I was involved in (high-quality pigs reared not in iron cages but in open pens on deep straw) ceased to be profitable.

In December 1990 I decided to cut my losses and get out – a worrying and stressful decision and a personal milestone for Cris and myself. The future looked bleak. I was a farmer and Cris was also born on a farm, so what else could I do? My heart was in my boots. By chance, the answer came out of the blue not long afterwards. I was often in the company of an RSPCA Chief Superintendent and chatting one day about farming problems, he said, "Well, George, whatever happens, you are never going to be short of work with the knowledge you have of the countryside, wildlife and caring for animals. Why not set yourself up as a badger consultant?" I said, "Don't be such a fool. Who on earth is going to employ me for work like that? I might get one day a month." "Think about it," he said, "you've got a lot to offer. Besides, what else could you do half as well? And what else would you enjoy doing half as much?"

So, I thought about it and decided it was worth a try. He could see the possibilities and perhaps he was right and I had nothing to lose. So, I took the plunge. My early days as a consultant were anxious ones. How much work would there be? What type of problems would come my way? My first job started when I received a telephone call from a well-known construction company carrying out redevelopment work in Birmingham. They had come across a series of large holes in a derelict railway siding and not knowing what to do, they had rung the RSPCA and spoken to Chief Superintendent Austin. He asked them if they had injured any animals and they said no, so he said, "This is nothing to do with us, but I know a man who can help you," and so he gave them my telephone number.

Shortly afterwards, the British Waterways Board asked me for help where badgers had been digging into the canal embankments. Then I had one or two cases where badgers were undermining roads. Next, a large national house building company contacted me, followed by many others including a quarry company. That led to a lot of survey work and my work seemed to mushroom. I was called in by the Highways Agency and Glamorgan County Council, again for work to do with new roads. Developers also by now were beginning to contact me, as they had badgers on their sites. Unfortunately, on most of the sites, they were calling me in after the work had started. It has taken a number of years to change that, so that I am now usually called in prior to construction starting, often two years in advance, at the very early stages of planning.

All of my work has come through word of mouth and personal recommendation – I find that heartening. Working seven days a week has always been my lifestyle, so it does not present a problem when the demand for my services results in a 70–80 hour week and in fact, I offer a 24-hour, 7-days-a-week service. I hope and believe that I do the job well, sensibly and professionally; protecting the badger in the face of inevitable change and helping well-motivated developers honour their obligations. Calling me in to sort out a problem costs developers money, but the other side of that coin is that bad advice can cost them thousands of pounds more and indirectly it can cause badgers all sorts of avoidable long-term problems. What is most important is that the advice must be independent, objective and free of bias. The onus is on the consultant to get it right, as he is financially liable if the wrong advice is given.

As luck would have it, my decision to get out of farming roughly coincided with the Protection of Badgers Act 1992 coming into force. The Protection of Badgers Act 1992 is an Act to consolidate the Badgers Act 1973, the Badgers Act 1991 and the Badgers (Further Protection) Act 1991. This law created a need for people who knew the ways of badgers, where they lived, fed and bred, and how best their needs could be accommodated in the face of relentless change: constant development in towns, villages and on the land.

That, basically, is how I made the change. Thanks to a combination of personal and national factors, George E. Pearce the volunteer, the enthusiast, became George E. Pearce, full-time independent badger consultant, beholden to no-one, but commissioned by developers, architects, ecologists, public utilities, construction companies and many more to carry out field surveys, give advice and construct practical solutions to 'badger problems'. I put that phrase in quotes because it depends, as they say, where you are coming from. Is it the badger that is the problem or is it the development?

Marooned: a bundle of badgers

Before I tell you about some of my experiences working virtually full-time with badgers, let me take you back to 1993 and to an incident that will live with me forever. I feel very privileged to have been part of it, for it really was extraordinary, as I hope you will agree.

Where the River Severn and the River Vyrnwy meet, close to where I live, their banks were overflowing after days of rain, and the local floodplain, the first piece of low-lying land that the gushing torrents of water from the hills of mid Wales can spill into, was around 90% full. It was quite a sight – floodwaters covered an area approximately fourteen miles long by three miles wide. I had often wondered what happened to ground-living mammals on this floodplain, where they took refuge and whether they always survived, for in wet years some would have to leave their homes and find alternative shelter five or six times a year.

It was early December and about mid-afternoon when I received a telephone call from the RSPCA. They asked if I could go to a farm close to the River Vyrnwy, as they had received a report that some badgers were marooned on what had become an island. The RSPCA couldn't get there themselves as they were busy rescuing sheep from the flood on another farm at Machynlleth in west Wales. I said, of course I would go, knowing that because of the floods, I would have to make many detours to reach the area.

The area is one I know well. At that time, it had 12 main badger setts on the land, which by now was under water, and I recalled that one of the setts was close to a hollow tree that the badgers headed for in floods. As I reached the farm, the floodwaters were still rising and I wondered whether, in two or three hours' time, I would be able to make it back home. The farmer met me and we made our way across waterlogged fields to the point where floodwaters were spilling over the local flood defence system, called an 'argae' in this part of the world. The top of the argae was three metres above the land that normally harboured the badgers, so we knew that in front of us was a huge wall of water held back only by the argae, which was already leaking like a sieve, with water spouting everywhere. On reflection, it was a potentially dangerous situation for us, let alone for the badgers, but at the time our concern was for the animals rather than for ourselves. The badgers were marooned on a spit of land no more than 80 m by 30 m, and which was just 20 cm above the floodwaters. Earlier that morning, the farmer had counted 11 of them and concerned for their safety, he had then called the RSPCA. I borrowed a pair of chest waders from National River Authority (NRA) staff, who were also at the scene, desperately trying to plug leaks in the flood defences. We knew that this was a truly exceptional flood, the worst in 50 years, and I was grateful that the farmer, who knew every inch of the land, was on hand to guide me to the shallowest areas as I made my way slowly towards the island.

As I neared this tiny spit of land, I saw something that was almost beyond belief. There were 11 badgers, two huddled at the foot of an elder tree, four were some 60 cm above ground, wedged between a wire fence and the tree trunk, and two more were up in the canopy of the tree some 2.5 m above the ground. Just a metre or so away were three more, in the bottom of a hedge. Oddly, all seemed very relaxed; in fact, so far as I could tell, most of them (including those balanced rather precariously up in the tree) were asleep! Astonished, I waded back and talked the situation over with the farmer and with a wildlife officer for the NRA. I also rang a local vet and discussed the matter with him.

We were concerned, which was more than you could say for the badgers, snoozing away in their unaccustomed makeshift home. The weather was mild, it was already late afternoon, and darkness was looming. We felt it was best to leave them where they were as they would probably swim to dry land and make their way to other setts on higher ground. As we made our way along the top of the argae, back to the farm, it was now pitch black. We had no torches, but we picked our way carefully through heavy waterlogged clay, doing our best to avoid water-filled hollows. There was water all around us and I was becoming increasingly concerned as to how I would get home. I had only a small car and the lanes were very narrow, undulating and awash, but I had to try. The first mile was a nightmare. In some places the water was level with the bottom of the car and it was the devil's own job to keep away from the verges, which I knew were bordered by deep ditches. Somehow, I made it without incident and, as I tumbled into bed, I wondered what the next day would bring for me and for the badgers.

The following day was bright and sunny and as I arrived at the flood banks there was no sign of the badgers on the island. I felt relieved: they had probably made it to higher ground during the night. The NRA staff had also arrived and once more I borrowed a pair of chest waders and pushed my way through the floodwaters, which were over a metre deep. As I got closer to the elder tree, I was greeted by a sight that will live forever in my memory. In front of me was a huge ball of badgers, half a metre above ground level, wedged between the wire fence and the tree trunk. Amazingly, they weren't in any distress – quite the contrary, they were asleep, their breathing synchronised! I had never seen a sight like it, nor read of anything remotely similar. I assumed there were nine badgers in the ball, as there were two fast asleep on a limb in the treetop. I stopped and pondered what to do. The weather was mild and the badgers were, to say the least, laid back and showed no hint of distress. Badgers under stress emit a strong smell from their scent glands and I could smell nothing amiss. They were sleeping normally, and one could almost hazard a guess that they were enjoying the novelty of being flooded out of their homes.

The following day I was back once more. The flood levels had fallen and there were no NRA staff, but I should be able (or so I thought) to wade across without much difficulty in my wellies. Fat chance! Soon they were full of icy-cold, crystal-clear Welsh water, and in no time I was up to my waist in water. In conditions like that you don't hang about, so I waded past the point which I knew contained one of their setts and quickly reached the island. The badgers were in exactly the same position as before, but I was concerned as the weather had changed and it was now very cold. How much longer could they survive like that, I wondered? I asked the NRA to arrange for the farmer to take 20 bales of straw across to the island in his boat, as these would make a shelter of sorts for the badgers. I also took a bag of dry dog food, as they would need something to eat, some source of energy, for they had already spent a considerable time above ground. I left them with the food and shelter and headed home, hoping that they would find both the food and the straw and make good use of them in my absence.

On Day 4 I approached the island very cautiously, expecting the badgers to be enjoying the warmth of the bales. To my surprise, they had chosen to sleep up in the tree and so far as I could tell, hadn't made use of the straw, although they had eaten a large quantity of dog food. In fact, it looked as though a herd of pigs had hit the island, as every square metre of ground had been foraged and I have never seen so many badger paw prints in such a small area.

Day 5 came and went, and again no change. I wrote in my diary, "These badgers really do have a mind of their own."

Day 6. I scanned the island – no badgers in sight. I approached carefully, fully expecting to find them buried in the straw. I removed one bale cautiously. Nothing. No badgers. No disturbance. They had at last left the island. I looked around and found paw prints at the water's edge. These showed they had not left in the direction I had arrived from, but had chosen a more direct route to a series of large clumps of overgrown hedgerows, which meant a swim of some 90–100 m, through floodwater up to 3 m deep in parts. I followed their route, reached the hedges and from there it was the easiest bit of badger tracking I have ever experienced. There was mud, mud everywhere and lots of large prints. The badgers had made 14 attempts to excavate a new sett in double quick time along a total length of about 400 m of hedgerow. Five of the setts had struck water as the badgers dug down and so had to be abandoned, but the other nine were relatively dry. I realised that they had chosen ground where their tunnels were still about a metre above the water table. They didn't need me any more, so I moved off very relieved and pleased that they had survived their flood ordeal.

I was keen to keep watch on these badgers, so I returned every two weeks to monitor and note what had happened. Throughout January, there was little obvious activity. In February, things changed and large areas of grassland were dug up as they searched for earthworms. In March, much to my surprise, they were back in their old sett – the one that had been so totally submerged. I thought it would take until July at least before it was dry enough for them to re-occupy it. A few months later, I paid another visit and noted that at least 18 of the 33 entrances were active and I was interested to see that their makeshift straw-bale sett was gradually being dismantled and taken under ground. It was a rare opportunity for them to have such a lot of bedding material available and they weren't going to miss out. By April, all the straw had disappeared and so, too, had the string that held the bales together. The sett was a hive of activity and in May I saw three cubs at the main sett. By June, of the 14 setts dug on that wet and wintry day, 9 were still in use. In July there was no change, but by the following month two more setts were inactive.

One year after the flood disaster, only two of the emergency setts remained in use. The badgers had no doubt chosen to move to other safer, drier areas, which was no surprise to me. What mattered was that they had survived; they had encountered a problem and resolved it in their own way (admittedly with some help) and life had returned to normal for them. As for me, I had helped in a small way and my reward was their survival and a memory that would stay with me forever.

Looking back on this episode, I wonder whether there was some inborn instinct, a reminder of the days, millions of years ago, when their ancestors were to be found in the canopies of forests, that had prompted those badgers to head for the trees and to comparative safety at a time of crisis. Who knows? Whatever the answer, this incident solved a puzzle that had stayed with me since my childhood days. Part of our farm was on a floodplain and every flood brought debris that piled up in a line much as it does on a beach at the top of the tideline. We had bottles and tins by the score, wood of all shapes and sizes, fencing posts and even complete farm gates, for in those days farm gates were made of wood. But there was one thing I couldn't understand: what happened to the moles, voles, rabbits, badgers and foxes that I knew abounded on the floodplain? We never ever found a carcass of any one of them. I had found no answer until those fateful floods of 1993. What the badgers had done is what most of the other animals I worried about would have done over the years: they climbed to safety or escaped, in good time, to higher ground.

Chapter Two

Badger biology

Scented signposts

The European (Eurasian) Badger (*Meles meles*) is a member of the weasel family (Mustelidae). Although classed as a carnivore, a flesh-eater, it is, in fact, a true omnivore and will eat almost anything edible.

Low-slung, short in the leg and long-bodied, characteristic of the weasel family, the badger is immensely powerful, with extremely muscular neck, chest and forelegs, and large, very strong claws. Instantly recognizable by its face, the head is white with two broad black stripes that run from the nostrils to the back of the neck. The ears are short and tipped with white, while the eyes look small compared with the size of the head and are set low down on its face.

The tough, white-tipped guard hairs, which cover the surface of the body, are white with a short, dark pigmented section close to the tip. Beneath the guard hairs lies the short, soft under-fur, which gives the badger a grey look, although in fact none of its hair is grey. In some individuals the black pigment is replaced by red, and the result is an erythristic or ginger-looking badger. Erythristic animals are commoner in some areas than others. For example, they occur reasonably frequently in north and east Shropshire, but rarely in the south or the west of the county. I have rehabilitated over 100 badgers and only five of these have been erythristic. Albino badgers also occur, but are even rarer and I have yet to see one in the wild.

Male badgers are known as boars, females as sows and the young are cubs. Over the years, I have weighed several hundred adult badgers killed on the roads in Shropshire and they averaged 12 kg (26 lb) for males, and 10.8 kg (24 lb) for females. The heaviest was a boar of 17 kg (37.5 lb). Body weight varies, of course, depending on the time of the year. Badgers do not hibernate during winter, but they do become very lethargic, feeding less often and for shorter periods. They prepare for the colder months by feeding greedily throughout the autumn and rely, in part, on the fat reserves put on during this period of abundant food to get them through the winter until spring. The result of this annual feeding pattern is that they are at their heaviest in late autumn and early winter and lightest in early spring, especially after a hard winter.

Badgers occasionally venture out in broad daylight, but normally emerge from their setts in the late evening and throughout the night. Their setts can be anything from simple, infrequently used, 'outliers' – often just a short tunnel ending in a single chamber – to centuries-old, surprisingly complex breeding setts with long runs of tunnels, several chambers and as many as 150 entrances.

Like their cousins – the otters, wolverines, pine martens, polecats, mink, ferrets, stoats and weasels – badgers have a scent gland at the base of the tail. This produces a musty odour (hence the family name Mustelidae). As they head off through woods, fields and along hedgerows to their favourite feeding grounds, they follow the same routes time and again, scent-marking as they go. This might not seem to be of any real significance, but to the badger and badger-watcher alike it is vital.

The badger's world is one dominated by smells and sounds. Its eyesight is poor, which is perhaps what you would expect of an animal that spends at least two-thirds of its time underground in dark tunnels and most of its time above ground in darkness. But its sense of smell is so acute that it is difficult for us to comprehend. Any scientist will tell you that, as humans, our sense of smell is remarkable and almost impossible to match, even with the most sophisticated modern laboratory equipment. Yet a badger's sense of smell is believed to be about 800 times better than ours – hard to comprehend! Their hearing is also very sharp, good enough to pick up the merest whisper or the faintest rustle of fabric, important to remember for anyone hoping to sit undetected near a sett.

When it leaves the sett to search for food, an adult badger normally follows well-defined tracks. Many of these will have been there for generations, even for centuries. Often, badger tracks have become our own footpaths, so it is not unusual for badgers and people to meet face-to-face at night, the one trundling out on its way to forage, the other wandering back after a late evening stroll. Eyesight and familiar landmarks will guide the human wayfarer, while scent marks act like signposts for the badger. Long-lasting and distinctive, such scent-marks also serve as warnings to other badgers, especially would-be intruders. In effect, they are saying, "Keep out, this is my land!"

I once saw a boar badger, close to its own sett, encounter what I can only assume was the latrine of a rival boar, possibly an outsider, a badger from another social group. I watched as it approached the offending dung with great caution, bristles up. It paused for some time, plunged its paw in and scraped out the contents of the latrine, dung flying in all directions. Then it enlarged the hole, deposited its own dung and moved off into the night, apparently content with a job well done.

Much is made, incidentally, of the assertion that badgers are very territorial and that boars, especially, will fight other males that intrude on to their territory and into their social group. I am sure it happens, but am not convinced it is as common as some wildlife observers suggest. Territorial defence is not that important to foragers such as badgers: however, defence of the social hierarchy is. It is the natural instinct of the dominant male to strive to pass on its genes and just as mature stags clash on the hillsides during rutting, so male badgers engage in horrific fights, both above and below ground, for dominance within their social groups.

Over 50% of the badgers I have received for rehabilitation have had fight injuries: bite marks on the neck, ears torn or bitten off or large areas of hair and flesh missing from the rump. These injuries are consistent with those of animals that have been fighting in a confined space, such as in an underground tunnel system. The front-end injuries are the result of head-to-head clashes, those on the rump having occurred when the aggressor attacks from the rear. Careful examination has shown that most of the bites have been inflicted over many weeks. That simply would not happen, in my opinion, if the disputes were territorial and occurring mostly in the open. By way of further evidence, I have often heard underground scurrying, snarling and panting sounds, and animals gasping for breath, followed, not long afterwards, by a badger bolting from an entrance, wet with sweat and saliva.

What's on the menu?

It is surprising how many people out for a walk in the countryside fail to see what is happening all around them but, for those who are observant enough, there are plenty of clues confirming that badgers are about – claw marks on rotting trees; dead wood that has been torn apart in their search for woodlice, grubs and larvae; snuffle marks where vegetation has been rooted up as they forage for titbits; straw flattened and twisted where they have consumed ripening grain, often already damaged by heavy rain; divots of torn-up grass; hand-sized holes scarred by powerful claws; and overturned cowpats (a reliable source of earthworms). Even more common are the tracks they make, and I deal with those in more detail in the next section.

Badgers, like foxes, have territories, or ranges as I prefer to call them, most of which contain a wide variety of food. Ideal badger habitat is undulating pastureland – livestock country – dotted with woods, hedgerows and copses. The trees and hedges provide both food and the cover the badgers prefer as they emerge at night and the permanent pastureland yields huge quantities of earthworms, the mainstay of the badger's diet. They also eat slugs, insects, small mammals, eggs of ground-nesting birds and carrion such as road-casualty rabbits, woodpigeons that have been shot and sheep carcasses that have remained unburied. Badgers also consume green plants, roots and bulbs, as well as fruits such as blackberries, elderberries, apples, pears, plums, damsons, nuts and cereals, especially maize. More and more farmers are using maize for silage as winter feed for cattle, as it ripens in autumn and is harvested in October and early November. That makes it a crop of tremendous value to wildlife, particularly to badgers, which need to lay down fat reserves to help see them through the winter. Badgers do, in fact, cause considerable damage to maize, flattening large areas and only eating one side of the cob, which is a considerable waste and annoys farmers. Badgers also have a great liking for wasp and bee nests, which they dig out with great speed and apparently without much discomfort to themselves.

In arable areas, many miles of hedgerows have been removed during the past 40 years. This has been very detrimental to badgers, as 80% of setts are found in hedgerows. Hedgerows have not been removed to the same degree on livestock farms. Such country then, is ideal for badgers; they like to build their setts in the hedgerows and the adjacent pasture provides an abundance of earthworms. The relatively high concentration of badgers in livestock areas leads many farmers to believe there has been a large increase in badger numbers across the country, which is not actually the case.

Observing a sett in April that had to be removed for a new bypass, my son Tristam and I noted the badger and fox tracks that crossed the very busy road that the bypass was replacing. We decided to conduct our observations from the car, parked in a lay-by, for the next few nights, in two-hour sessions, which would help us to understand why there were so many badger setts. There were 18 in all, but only one was active on the site. However, there were a large number of fox tracks and scats throughout the site.

First night: 7:00 pm to 9:00 pm

The traffic volume was very high. At 7:20 pm, a dog fox emerged from the site and very casually searched for food in the grass verge and on the road. Each time a car passed, it slipped through the hedge for safety and when there was a gap in the traffic, it walked casually back to his foraging. This lasted for 45 minutes, then it went back to the site.

Second night: 9:00 pm to 11:00 pm

A few minutes after arriving, a badger emerged through the hedge from the sett and foraged along the grass verge and shallow ditch, completely oblivious of the traffic. After a while, it ambled across the road into a large grass field, at which time the dog fox that we had seen the previous evening came on to the road, together with a vixen and three cubs. They foraged along the grass verge and roadside for a few minutes and then crossed the road a few seconds before a car came speeding by. Then all seemed quiet until the badger decided to go back to its sett area.

Third night: 11:00 pm to 1:00 am

On arrival, we were aware of the low volume of traffic – just two cars passed by in the first 15 minutes. Then the vixen came through the hedge from the field with her three cubs and very casually they all lay down on the road to rest, soon to be joined by the dog fox. After 15 minutes or so, they all got up and foraged on the hard road. We were at a loss as to what they were eating. This went on for a few minutes until they were disturbed by a car. When the car had gone, we got out of our car and with the help of a torch, searched the road to see what they had been feeding on. To our surprise, there were hundreds of black beetles, which were, in turn, feeding on twigs that had been broken to a pulp by the traffic. I have been aware for a long time that badgers and foxes consume a large number of these beetles in and around badger setts, but was not aware that they were in such large numbers on our roads.

Fourth night: 1:00 am to 3:00 am

It was very cold and frosty when we arrived and as expected, the road was very quiet. An hour had gone by when we saw the same young boar badger we had seen on the second night. It came through the hedge from the grass field and spent a considerable time foraging, first along the grass verge and then on the road to feed on the beetles. Judging by the speed at which it was eating, there was a good supply of beetles. How often, as we drive along the country roads, or indeed along the roads in our towns and cities, have we disturbed badgers and foxes, forcing them out of the way a few seconds before we pass by?

Most of the food that comes the badger's way is nature's gift. But, if man (in the guise of a badger-watcher) lends a hand, then 'Brock' is all too ready to take what is on offer. Adult badgers and cubs will tuck into peanuts, peanut-butter sandwiches, jam butties, syrup and honey, sultanas and raisins, without pausing for a second to stop and wonder where this bounty comes from. That said, it is worth pointing out that people who feed badgers (and I'm not against that, in principle) are creating artificial conditions. What they see when they place peanuts outside a sett is not what would happen in normal conditions. They see the badgers close up and for longer, but they miss out, in the sense that they are not observing natural behaviour. They miss watching the badgers being very cautious when emerging, the grooming and scent-marking on other members of the group and on objects around the sett.

Follow the tracks

I use a simple rule-of-thumb while carrying out wildlife survey work: the higher the number of tracks and the more wear and tear they show, the greater the number of animals there are in the immediate vicinity. Heavily vegetated areas where farm livestock have no access can often be very revealing. Some show virtually no disturbance, no signs of use. Others are like road maps – look closely and you will see some tracks so obvious and well used that they equate to motorways and A-roads. Others not so well used are the B-roads, while others, fainter still and less well marked, are the unclassified roads and bridle paths. Study them, and you will gradually be able to pick out characteristics that give you a strong clue as to which animals are currently using them. There will be prints, droppings, signs of foraging, leaves that have been nibbled, trees or roots that have been scratched or bitten, as well as fragments or remains of partly eaten food. Find a well-used badger track and with luck and patience, you can learn something new.

Main tracks measure about 10–15 cm (4–6 in) in width and typically they are similar to public footpaths insofar as they do not turn and twist rapidly. Follow them carefully and you will spot that they gradually change in size and appearance. If they become smaller and more difficult to follow, you are heading away from the sett, but if they become wider and the surface gets shinier, then you are walking towards a sett. Smooth, well-worn, almost grass-less paths close to setts indicate lots of activity and that generally means that the sett holds several badgers, perhaps as many as six or eight, though it is surprising how much wear and tear even a single badger makes at a sett.

Badger tracks can alter in shape and size within the space of 50 m. At other times, depending on the number of badgers using them, the point at which you join the tracks and the type of soil and vegetation they are passing through, you can follow distinct paths for a mile or more. These invariably lead to important feeding and foraging areas. Some temporarily disappear if they cross land that has been ploughed, or where large building projects such as motorways and bypasses are being built but, rest assured, this has no effect on the badgers and on their ability to find their way. Their scent trails, laid down over weeks, months and years, are so strong and the badger's sense of smell so acute, that they continue along exactly the same line as if the track had remained untouched. I have witnessed this many times.

Disused badger paths are also readily identifiable. Vegetation fails to grow where the soil is well worn and compacted, but the tracks do moss over. Do not expect to see thick moss, look for a thin coating, almost a shiny green film. This tells you that heavy paws are no longer pounding that path and wearing it smooth.

All too often, while studying badgers and more recently when carrying out survey work, I have found several paths that meet at a particular point, often at a small scratching tree, usually elder. Indentations close by, too regular to be natural, tell me that this is the site of a sett that has been illegally destroyed. It spoils my day.

Let us touch briefly on those prints I mentioned. In wet weather or in damp conditions and especially in clays, loams and sandy soils, it may be possible to spot the large, distinctive prints of the badger. Look for five toes, but do not be surprised if you only see four, and an unusually wide pad. When I am giving talks or field-craft lessons to children, I usually tell them to close their fist and press it, palm down, into sand or soft soil. That leaves a print not unlike the badger's paw print. Soft, damp conditions or white frost are ideal for more field-craft detective work. Foxes, rabbits, squirrels, hedgehogs and deer all leave distinct imprints and fresh snow is an absolute delight for the would-be tracker. The dainty straight lines left by a fox, the long hind feet of the rabbit as it hops around, the clear hoof prints of deer, all stand out in crisp snow. You will also see where rodents have been scurrying, dragging their tails through snow that is often much deeper than they are tall. In milder weather, with more yielding conditions underfoot, you will find, with experience, that foxes create two types of track – one while hunting, the other while foraging. The hunting track usually runs in a straight line parallel to a hedge, some 3 to 4 m from the centre of the hedge. Foraging tracks are shorter, less well defined and they zigzag, a sign that the fox has been scenting much smaller prey like beetles and grubs. The longer the grass, the greater the imprint left by a fox's paw. Where rabbits are plentiful, you will find many small tracks, with lots of directional changes and plenty of tell-tale, raisin-sized droppings. As a rule-of-thumb do not expect to be able to pick out the tracks of stoats, weasels and polecats. They are difficult to spot, and even more difficult to identify with any certainty, unless you are lucky enough to find breeding sites.

Stiles and fences are always worth close examination if you are looking for signs of badgers and foxes.

At a stile, I look at the bottom bar. Hair snagged on the underside (which may also appear shiny from regular contact) will usually be from badgers as they go under the bar, foxes tending to go over. Where paths go through hedges and fences, look for tufts of hair caught on wire, brambles, thorns and the like. You will find that the underside of the thorns on some brambles have been broken off by constant contact. Badger dorsal hair is white for approximately half its length from the base. It then has a section of black pigment and finally is tipped with white. Unlike fox hair, badger hair is coarse and when rolled between finger and thumb, it rotates unevenly.

Finding and identifying animal hair is a useful skill for anyone interested in wildlife. For me, it is part of my job, and I always carry a few small plastic bags to collect samples in case I have to provide proof that badgers are present. Let me give you just one example. I was appearing as an expert witness for the prosecution in a badger-digging case and was questioned by the defendants' solicitor about two paths leading from the sett in question, a sett they were trying to prove was not used by badgers. One path led in a northerly direction, the other went east across an open field containing cattle and sheep.

He asked if I had seen cattle footprints on these paths and I said ,"Yes".

"Did you see sheep footprints on the paths?" and I said, "Yes".

"Did you see badger footprints on the paths?" and I replied, "No".

"Did you see cattle dung on the paths?" I said, "Yes".

"Did you see sheep droppings on the paths?" Again, I replied, "Yes".

"Did you see badger droppings on these paths?" I said, "No". Here, there was a long, pregnant pause.

"And you expect the court to believe that badgers were using these footpaths?"

"Yes", I said. "Where the paths reach the wire fence, the cattle and sheep turn left and the badgers go straight underneath." I then referred the court to a photograph I had taken at the spot, showing badger hair caught on the lower strand of the wire fence and a latrine close by.

The case was successful and the two defendants were both jailed.

Clues in the latrines

Latrines are another indicator of badger presence and very distinctive they are too. Badgers usually defecate in small, cone-shaped depressions, generally some 5–20 cm deep, which they dig themselves and, unlike cats, they leave them uncovered after use. Occasionally, the latrines are larger: I have found some up to 35 cm deep, often under brambles or other overhanging bushes. In dry conditions adult badgers will defecate on top of the ground – it is quite common for cubs to do this in their first few months above ground and small heaps of faecal deposits near a sett in early summer are a good indication that it may well be a breeding sett. In spring, it is also common to find latrines close to, or even in, the entrance of breeding setts.

The commonest places to find latrines are where badgers were feeding the previous night. They are also found close to badger paths, by boundary fences and near to fruit-bearing trees during the fruiting season. I have also noticed that, in winter, the colder the weather, the closer the latrines are likely to be to occupied setts. This is especially noticeable when temperatures drop below freezing. In urban areas, latrines are normally much closer to the setts than they are in rural areas. A latrine will often be used a number of times, but may then be abandoned for up to a year before being re-dug and used again as the badger revisits that particular feeding ground for food that has come back in season in the enriched soil.

When you find a latrine, it is clear whether the droppings are fresh or old and a cursory glance will often give some clue as to what the badgers have been eating. If the droppings are very loose and muddy, that is a sign that they have been feeding primarily on worms. In bluebell woods, latrines may contain white fragments of corms. Remains of black beetles, plum stones, apple skins, bits of maize, blackberry seeds and particles of all types of fruit and vegetables are also often evident, as are indications that carrion has been consumed.

Closer examination tells you even more if you have the time and the stomach for it! Over the years, I have collected hundreds of badger and fox droppings for detailed examination at home, though unfortunately, as my wife, Cris, will confirm, I also have a habit of leaving them overnight on the car floor in a plastic bag or even between a couple of dock leaves, where Cris finds them next day. I do not recommend that as a way of impressing your wife! To examine the droppings, I put the material into a fine sieve kept especially for that purpose – a flour sieve is perfect. Place a white container under the tap and let the soft, muddy material run through into the container beneath. The solid particles then settle in the bottom of the sieve, and most of these can be identified. Not the most glamorous of operations, but a very good way of learning more about what both badgers and foxes have been eating. I am sure that if they handed out PhDs for the number of years of studying farm and wild animal dung, then I would have received one years ago. You can also learn something about the animal's health from such examinations.

Very revealing, but even less pleasant for the squeamish, are autopsies, especially when examining the stomach contents of badgers recently killed in road accidents. I have done that on numerous occasions before burying the remains and have learned a great deal. But I must add that, under the Protection of Badgers Act 1992, you need permission to take a carcass away. Think about it and you will soon realise why. Once you have picked up a carcass, you are 'in possession' of a badger and with badger-baiting still rife, a police officer might be very keen to ask you a few difficult questions.

I would add one more comment about latrines. A great deal has been written about them and much has been made of the assertion that very often they mark the boundary between one social group of badgers and another. In areas of high or unusually high badger density, this may well be so, but in my experience, in areas with average levels of badger density, they are most usually found at feeding sites and are simply an indicator of where badgers were feeding the previous night.

Spoil heaps: mountains of information

Spoil heaps are the badger's equivalent of builders' rubble. Some are so large that they can be spotted from a distance of a quarter of a mile or so. The fresher they are, the more obvious they tend to be. Spoil is what the badgers excavate as they create the setts in which they live, rest and breed. Some setts are in the middle of a field. More often, they are in hedgerows or under trees in woodland, usually within 100 m of the edge of the wood. Spoil heaps are a mountain of information. Whilst badgers are excavating in damp clay soil, bits roll back down into the tunnel, making small round balls that vary from the size of a marble to the size of a tennis ball and badger hair can often be found entwined in these. If you stop to examine the surface, do so with care, and remember that spoil heaps form part of a sett and that setts are protected by law.

The size of the spoil heaps will normally reflect the size and age of the sett, and the ground conditions. Even the badger has a tough job tunnelling into solid chalk and flint or into stony terrain, so setts and spoil heaps in these kinds of substrates tend to be small. The constant passage of livestock, or the use of heavy farm machinery, will also damage, flatten or distort the shape and size of the heap. The largest single spoil heap I have ever examined was in undisturbed woodland – it measured 4.5 m long, 3 m wide and 2.1 m high.

If soil and weather conditions are right, the spoil heap is a wonderful place in which to find prints. All types of burrowing animals live quite happily in setts occupied by badgers. Waiting by a sett to see badgers emerge, I have observed, at various times and locations, foxes, rabbits, woodmice, rats, weasels, polecats, mink and, would you believe it, even cats, emerge before the badgers. It is common to find rabbit droppings in and around the entrances to setts and it is often assumed quite wrongly that this is a sign that only rabbits are living there. Do not be misled, but look for the other indicators of badger presence – paw prints, badger hair, fresh vegetation taken into the tunnel system and discarded vegetation previously used for bedding material. Unlike foxes, which occupy bare earths, badgers like to line some of their sleeping and nursery chambers with vegetation they can gather locally. If they live in or near the edge of a town, they will often even drag in plastic shopping bags and crisp packets. In various setts I have found 10 m of rope, electricity cable, large plastic sheets, two doormats and a mop with a 1.3 m handle.

Many observers have commented that, from time to time, badgers also pull bedding out of the sett and leave it near the entrance to air, a belief founded on the fact that small piles of used bedding material are found on spoil heaps. I believe this conclusion to be misguided. Badgers certainly drag quantities of bedding out, but my belief is that it happens by chance. Badgers are continually extending and cleaning out tunnels after roof collapses and the bedding found on spoil heaps is, in my opinion, material which has become mixed with newly dug soil or unwanted debris they want to remove. In effect it has been 'thrown out with the dishwater'. But once the internal spring-cleaning or sett extension work has been completed, reusable bedding will be dragged back in and used again. Argue as we might about why so much vegetation ends up in spoil heaps, especially at main setts, one thing is certain: the presence of so much decaying organic material explains why large numbers of beetles are found in spoil. Over the years, I have seen foxes and badgers spend a considerable time on and around spoil heaps, tucking into food of some kind and it was not until I began consultancy work that I found out what was attracting them. Working under licence on a sett that had caused road subsidence, I had to remove a large spoil heap. On dissecting this carefully, I saw that, from all directions, beetle burrows penetrated to the heart of the heap, and there I found literally hundreds of beetles 'sleeping' in their version of a large nest. Both foxes and badgers regard beetles as very tasty morsels and their remains are often found in their droppings.

Anyone who spends a lot of time close to setts, watching badgers emerging, returning, scampering and scratching will have tales to tell. Let me recount just one tale that underlines how preoccupied they become when excavating tunnel systems.

Some years ago, I arrived to watch a favourite sett and found an old buck rabbit sitting on the bank about 2 m above the sett entrance. We gazed at each other for about half an hour as the sun disappeared behind the Berwyn Mountains in North Wales and the light began to fade. It was then that a warm, musky scent filled the air. Moments later, a pile of soil appeared at the sett entrance, followed by a badger using its backside like a bulldozer as it pushed its way out backwards, a pile of soil clasped between chest and front legs, manoeuvring as much as it could to the edge of the spoil heap. It stopped, shook itself thoroughly, cocked its head, first towards the rabbit, then suddenly towards me as it picked up my scent, paused for a moment and then hurried back into its sett. And that, I thought, was that, my viewing for the night was over. Not a bit of it. A few minutes later, out came more soil, followed by the badger with more soil to dump. Once more it shook itself, turned to me and then to the rabbit. Chuckling to myself, I said in my normal tone of voice, "My word, you are busy tonight, mate." Startled, it froze, with one paw in mid-air, held that position for around 10 seconds, then turned and went back into the sett. No doubt about it this time I thought, the action is over; time to go home. But I waited just in case and once again out it came with more soil. Same sequence: it looked at the rabbit, glanced towards me and disappeared once more. Next time it appeared, I offered another comment "That looks like hard work. How much longer will you be?" Again, another thoughtful glance in my direction, a long pause and once more it vanished. Fourteen times this sequence was repeated before finally it shook itself, scent-marked the spoil heap and casually wandered off into the night. Even now, it makes me smile. A buck rabbit, a badger and me, caught up in the sort of cameo that you would expect to see in an episode of *Last of the Summer Wine.*

Let me give you a few more pointers about setts and what to look for because, as I say, they are the best evidence you will find of badger activity. Most are exactly as they are described in many books: the entrance hole is slightly wider at the base than at the top, and the tunnel, for as far as the eye can see, is wider than it is high. At old-established setts, the amount of spoil is vast, quite literally tonnes of soil with discarded bedding, anything from dry grass, straw or leaves, even pine needles are to be found mixed in with the spoil. At newly dug setts, the shape of the spoil heap is reasonably distinctive: usually neat and tidy, unlike the spoil outside a fox's earth. Foxes tend to scatter soil, throwing it back in much the same way as a dog digs. Large stones and even small boulders with badger claw marks are also frequently present in the spoil at a sett. Only badgers could remove those. Examination of the loose earth reveals more clues: the distinctive hairs of a badger (see earlier section: *Follow the tracks*) and huge paw prints. The sides of the entrance are also often scored by what are clearly large, sharp claws so typical of a badger and 'scratch trees' (often elder) are frequently found close to well-established setts. The presence of wide, well-worn tracks leading away from the entrances is another indicator of badgers rather than foxes.

Some setts have oddly shaped entrances, this is the result of various influences like tree roots, immovable boulders, the effects of the weather or the looseness of the soil. In light soil, fox earths can look like setts and entrances to rabbit burrows can sometimes be as large as those of badger setts. By contrast, in difficult, stony or chalky terrain, setts can be quite modest, with a small entrance and short tunnels, with very little spoil.

In rabbit warrens and at the mouth of single rabbit burrows, unsurprisingly, you will find lots of rabbit droppings. But, just to muddy the picture, rabbit droppings are quite common, too, at sett entrances. Rabbits often wander in and out of sett entrances, preen on the spoil heap, or simply hop by *en route* to somewhere more interesting. From time to time, they also take advantage of the badger's hard work and live in part of a sett. Foxes do the same thing and usually badgers and foxes sharing the same sett ignore each other. A fox would be no match in a straight fight with an adult badger – but foxes will occasionally kill badger cubs and badgers will occasionally kill fox cubs, which partly explains why badger and fox cub skulls are commonly found in spoil heaps.

This habit of different animals sharing what is, after all, very desirable underground accommodation – large, dry tunnels with comfy chambers, some liberally lined with bedding and temperatures that are neither too hot in summer nor too cold in winter – means that sett entrances often offer a range of conflicting visual clues: fox hair, rabbit hair, badger hair, dog and cat hair, hair from cattle, strands of wool and facial hair from sheep, rabbit droppings, fox, polecat and stoat scats and several types of paw prints. Large, active setts with fresh spoil, lots of badger paw prints, nearby latrines and badger hair in the spoil are the easy ones to identify. The smaller, decaying setts with fallen leaves and other debris in the entrance and few other indicators of current use, or with a mixture of conflicting visual clues, are more difficult, and often it is necessary to monitor these by day and by night over an extended period. Straightforward night surveillance will sometimes provide the answer. But, if you have done any badger-watching, you will know that, although some setts are active every night in normal weather conditions, others are much more unpredictable and it is possible to go several nights without seeing a single badger emerge. So, night surveillance often has to be supported by entrance monitoring. Badgers are not exactly light of foot; they lumber along in their purposeful, short-sighted way on large paws, brushing aside bits of debris as they enter or leave a sett and leaves, sticks and other pieces of debris are constantly falling into a sett entrance.

Chapter Three

The world of the sett

Secret world of the sett

Setts come in all shapes and sizes and are excavated in all types of geological substrates; some are centuries old, but new ones are being built all the time. I know of one that appeared overnight; it was dug by a single badger and I was able to measure the length of the tunnel: 5 m! The badger did not take up residence immediately and continued to extend it every few nights. If the soil conditions favour digging and tunnelling, the badger's strength and persistence can have quite extraordinary results.

One sett proved quite an education. It was August and a local farmer called me in to ask what could be done with a sett that had appeared in the centre of a cornfield during the growing season. He did not want to damage it, or injure any badgers, so he had left an area of about 20 m by 30 m unharvested. He explained that, much as he, his wife and small sons enjoyed watching the badgers in another sett on the edge of a field, he couldn't afford to sacrifice such a large area and the sett out in the field would create problems when he ploughed and re-sowed in the autumn. Working under a Ministry of Agriculture, Fisheries and Food (MAFF, now DEFRA) licence, I placed one-way gates (similar to a one-way cat flap, only larger) on the four entrances and checked them daily for 21 days. Once I had proved that the badgers were no longer living in the sett, it was excavated by machine, starting first at the entrances and following the tunnels along to each of the chambers. In all, we found 12 chambers, four of which contained bedding material, and the tunnel system measured 80 m in length. An official from MAFF, incidentally the most knowledgeable wildlife expert I have ever met, was present during the excavation and when the farmer told us that the sett had not been there when he sprayed the crop in May, we both glanced at each other with a look that said the sett must have been there for years and the farmer had simply ploughed over it. Nothing more was said and we both left.

Exactly one year later, I received a telephone call from the farmer, saying the badgers were back in exactly the same spot in the centre of the field, which was a small area of sandy soil surrounded by heavy clay. I went out to look and could barely believe my eyes: this time there were eight entrances. We had to go through the same procedure again, closing the sett under licence. We unearthed ten chambers and I measured 110 m of tunnel length. Without doubt, the whole sett had been dug since May, which was less than three months earlier. I tell this story to underline what remarkable diggers badgers are, a fact confirmed by other setts that I have had to close, mostly because of imminent building development or threatened structural damage to property.

No two setts are the same and they vary tremendously in size and configuration: expressed in our terms some are tiny country cottages, others are grand mansions. The shortest tunnel I have come across (and I have examined over 1,500 setts) was just 1.1 m, while the greatest total length of tunnels in a single sett measured 260 m. Some setts have three storeys of tunnels and chambers. The size of tunnels within setts varies according to the type of soil and the age of the sett. My records show that the narrowest have averaged as little as 18 cm, the widest 30 cm. Heights have varied from 13 cm to 23 cm. Chambers also vary in size and shape, from 40 cm to 122 cm in length, 30 cm to 80 cm wide, and anything from 25 cm to 120 cm high. The greatest number of chambers in any single sett I have closed was 42 and of these, 18 contained bedding material.

This sett was in dangerously contaminated soil. Tunnels are dug mostly on strata joints and those in light soil and sand have the largest tunnel systems. Although setts in heavy clay are often quite large, those in chalk and flinty terrain are characterised by rather short tunnels with many entrances, while stony and coal-bearing areas usually have small, two- to three-entrance setts. The size of the sett cannot be judged by the number of entrances.

When working on a site half a mile from the centre of an industrial town, we moved the badgers from their natural sett to an artificial sett so that their original sett could be closed to allow new houses to be built. We had arranged for a large mechanical digger to excavate the entire tunnel system under our supervision and about 2 m into the sett, we started to find crisps packets and supermarket plastic bags. A little further into the sett, we found a large plastic sheet. We pulled and pulled at this sheet until it came free of the sett and when we spread it out, it measured 6 m × 6 m. We were still finding crisp packets and carrier bags in their hundreds and next we came across a piece of carpet 3 m × 4 m tucked into one of the chambers. We know that badgers collect bedding, but this quantity was so unusual. Still further on into the sett, we came across a partly worn coconut doormat, then a mop with a 1.2 m handle. We were now 12 m from the entrances and how on earth they managed to get it so far down the sett's winding tunnel system I will never know. Following tunnel after tunnel, chamber after chamber, still finding crisp packets and carrier bags galore, we then found a brand-new coconut doormat in a rather large chamber. I have learned to expect the unexpected where animals are concerned. Across the road from this sett was a row of typical northern industrial houses and one can imagine a householder accusing her neighbour of stealing her doormat, then having replaced it with a new one, only for that to go missing as well!

I have explained that some setts often have rabbit droppings at the entrance and that many are large enough to accommodate uninvited guests, including foxes. So, there are often signs in the spoil that make it difficult at times to decide whether the opening is the entrance to a badger sett, a fox earth or a rabbit burrow. Typically, the entrance to a badger sett is wider at the base than it is tall, but sett entrances, just like the setts themselves, vary a lot and you need to look beyond the entrance, about the length of your arm into the tunnel system, to get a true feel for the shape and the size of the tunnel as it disappears.

As you look, keep in your mind the body shape of the rabbit, the fox and the badger. Rabbits excavate a tunnel that fits their body size (which is not all that much bigger than a cricket ball in diameter). The fox, being a long-legged animal with a slim body, typically excavates a tall, narrow tunnel with very few bends and just one simple bare chamber at the end. Badgers, being very muscular and short-legged, dig a low, wide tunnel, which will have a bend or a fork less than 2 m from the entrance. Usually, a badger tunnel dips quite sharply away from the entrance.

Keep these simple guidelines in mind and you won't go far wrong if you are trying to determine whether you've found a rabbit burrow, a fox earth or a badger sett. Always allow for the fact that the entrance hole may be much larger, for all sorts of reasons, than the tunnel. Bear in mind, too, that the soil inside the entrance of an active sett will often be worn shiny smooth and that the badger has the strength as it tunnels to remove and drag out stones which would be much too heavy for foxes and rabbits to move (look closely at these stones for claw marks). If the spoil contains soft sandstone, chalk or clay, you may be able to spot badger claw marks in these. A fox is approximately half the weight of a badger and a lot longer in the leg. When foxes dig a tunnel, it is generally shorter, so the spoil heaps are smaller and because they have longer legs, they also throw the excavated soil over a wider area, creating a flatter mound than when badgers dig.

If you are trying to puzzle out how large the sett is and how many badgers could be inside, do not base your conclusions on the number of entrances, as they are not a reliable guide. The best indicator, as I have already said, of the numbers of badgers using a particular sett is the amount of wear on the paths leading to and from the sett. Do not assume, either, that simply because some of the entrance holes are blocked or crumbling, the sett is inactive. Time and again in discussions with farmers and badger groups and even in court cases where badger digging is alleged, I am faced with assertions that part of a sett is disused. Take an eight-entrance sett as an example. A cursory glance is often enough to tell you that some of the entrance holes are not being used: the sides and top may be crumbling and twigs, leaves and other debris may obscure some of the entrance holes. All this is sometimes taken as evidence that part of the sett is also disused, but none of the sett excavation work I have done supports that theory. Invariably, the entire sett is used and all the tunnels are clean and polished with wear except for a few centimetres close to those disused entrance holes. Why some of the entrances briefly fall out of favour I do not know, but what is important is that such setts may well be active despite those 'closed doors'. With the exception of two that I shall refer to later, every sett I have ever opened up has been clean and even when the tunnelling has extended for more than 200 m, the signs of regular, daily use are everywhere. The tunnels are worn clean, and may be spotless, so spotless that I have been tempted at times to look for the three-point sockets where they plug in the vacuum cleaner! But seriously, those ultra-clean tunnels have made me think and I believe I have unearthed an aspect of the badger's lifestyle that has not previously come to light. I shall tell you about that in the next chapter.

One other tip: in very cold weather, north- and east-facing entrances are often blocked by the badgers from the inside, using bedding material. I have also observed similar behaviour from badgers in captivity.

Tunnels: DIY larders

I mentioned earlier that my consultancy work has given me an intimate knowledge of setts and as a result I have developed a theory, which I hope is well worth explaining. So, here goes.

Years ago, when I spent so much of my spare time badger-watching, I'd often arrive at a sett up to a couple of hours or more before I expected them to emerge. On a summer's evening the badgers might come out at about 9.30 pm, but well before that, as I sat listening and watching, I often detected activity.

I am blessed with an exceptionally keen sense of smell and on a still, mild evening I frequently picked up warm, musky scents drifting out of the sett, accompanied by sounds which told me there was a lot of movement underground. At the time, it was an interesting observation, nothing more. Now, with the benefit of years of consultancy work, which has given me the chance to open up and minutely examine setts from which badgers have been excluded, I believe I can explain much of that underground activity and more importantly, why setts often have what seem to be excessively long tunnel systems.

It is my firm view that tunnels are not merely passageways between sleeping and breeding chambers – not simply routes between *A* and *B*. They are also used as larders, providing food, especially useful when the foraging grounds up above are baked dry and earthworms, the badger's favourite food, are difficult or impossible to find. I will develop this point a little later. Let us first look at some of the alternatives that have been put forward to explain the badger's propensity for tunnelling.

(a) Ventilation

The theory goes that tunnels are dug to provide the right level of ventilation to create air flows through the tunnel complex. To me, this does not add up or at best it is only part of the answer. Setts have too many tunnels with dead ends. One scientist who did a lot of very impressive work looking at this aspect of badger behaviour and the effect of air circulation in the tunnels, reached the very interesting conclusion that badgers need very little by way of fresh air because they can make do with very low levels of oxygen.

(b) Disease control

The suggestion is that a network of underground tunnels allows the badgers to isolate disease and keep it to one part of the sett, while they use the remainder. This is not practical, as I do not think it would work quickly or effectively enough: it would take months or even a year or more for the sett to be free of disease if left empty. It would make more sense, surely, if disease were a recurring problem, for badgers to use a number of small setts.

(c) To 'rest' part of the sett

This simply does not seem to happen and all the setts I have excavated have shown the same level of use throughout. To repeat, the tunnel systems invariably are scrupulously clean and obviously are in regular use.

(d) To avoid each other

Long lengths of tunnels might offer some temporary advantage, but I'm sure an aggressor would continue to pursue its victim irrespective of the size of the sett.

(e) Parasitic control

The thinking here is that badgers desert infected chambers and use others, leaving the parasites behind. The kind of parasitic infestation we are concerned about here does not work that way: the parasites live on the host and will therefore move with it wherever it goes. Some experimental monitoring work done with radio tracking devices found that badgers often slept in a chamber for two or three nights, before moving on. The reason, it was suggested, was the parasitic build-up, but again I think that the wrong conclusion has been reached. Badgers leave sleeping chambers when the bedding turns mouldy. Collected when damp, it then warms up, at some stage is allowed to go cold and as a result it turns mouldy. By contrast, bedding collected in dry periods will stay usable for a considerable time.

So, back to my belief that tunnels act as larders. Almost without exception I have found that the 'ceilings' of badger tunnels are honeycombed with the burrows of beetles and worms, and during sett closure work I have noticed that there are almost always large numbers of worms and beetles on the floor of the tunnels. They did not fall simply because we were excavating; rather we were seeing something, I am sure, that happens all the time. Charles Darwin observed that the type of food that an animal eats and the availability of that food, will determine the behaviour of that animal. I am sure that, over thousands of generations, badgers have adapted to the fact that tunnels can serve as larders and have altered their behaviour to their take advantage of food dropping into their sett. The mole lives all its life underground collecting food that drops into its tunnel system, so why not the badger? They both feed on worms. I cannot be sure, but my bet is that, between waking and emerging, badgers travel those tunnels several times, which explains the smell that comes and goes at sett entrances and the sounds which I (and no doubt lots of other badger-watchers) have heard from inside setts before the animals emerge.

There are other clues that point to tunnels as larders. For example, studies have shown that badgers, though they do not hibernate, can spend a long time underground without ever emerging for food. MAFF (now DEFRA) officers recorded one account of badgers staying underground for 28 days, yet when they emerged they were fit and well. Had they been raiding their own underground larder? Around 80% of the setts I have had to close and excavate had no badgers inside them at the time the one-way exclusion gates were put into position. Yet, the tunnels have invariably been clean and well used and at the tunnel entrances fresh badger paw prints were visible and well-worn paths led to the setts. These are all signs suggesting that, although the setts were not being lived in, they were being visited simply to obtain food, especially in hot, dry or frosty weather, when foraging above ground would be difficult.

Secondary or outlier setts provide the same kind of persuasive evidence. Smaller than main setts and usually some distance from them, they are used chiefly in the summer months. Badgers have often been observed entering such setts at night for short periods of an hour or so, then re-emerging to return to the main sett. They would not be visiting to have a nap in the middle of the night, so the most likely reason is that they are in the tunnel system foraging for earthworms.

The lifestyle of the earthworm offers another clue. Ask a gardener what happens to the worms in dry weather: they disappear. They have, in fact, gone much deeper down into the soil. When that happens in our open fields, hedgerows and woodlands, the worms are living at depths at which our badgers can ease them out of the tunnel walls or simply collect them as they drop to the tunnel floor. No animal does 'owt for nowt'. If it expends a lot of energy, then there must be a reward, a plus factor that makes the effort worthwhile.

Finally, while we are on the subject of tunnels, let us deal with a practical problem: dogs down setts. Small dogs, terriers especially, are full of energy and always up to mischief. It is in their nature to explore anything that smells interesting. Sometimes, they get over-inquisitive, disappear down a sett and refuse to come out. The owner then panics and calls the emergency services. The police, RSPCA or the fire service then turn up, sometimes all three, and as a result there is a great deal of commotion, and that makes the situation worse. The dog is frightened by all the strange sounds and scents, becomes even more nervous, and is then less likely to come out.

My advice is that the owner should ask everyone except

family members to keep away. Then, if it is practical, just two members of the family should stay at the sett and take it in turns to try, quietly and gently, to coax the animal out. It may take some time, up to 24 hours or more, so it might be sensible for each person to take shifts, say for approximately four hours each.

If you bring stools or cushions and food and drink, that makes the waiting more comfortable. It is essential to keep talking to the dog in a very relaxed voice. Read a book out loud if that helps to pass the time, just keep talking so that the dog hears and knows you are there. Anyone who is upset or panicking should be asked to stay away, as this could make the dog more anxious.

You might not think so, but there are very few places in a sett where a dog can become trapped. If it cannot go forward, it can always back out: four-legged animals are very adept at reversing out of tight situations. If the dog is down there a very long time, one possibility is that it has encountered a badger and there is a standoff, each refusing to move back. If you panic, it may too. So, try to calm it down. Keep talking. I can assure you it works. On numerous occasions, I have been asked to visit a sett by an over-anxious dog owner. I used to go, but I don't any more, I just pass on the advice over the phone. It is very rewarding later when the phone rings and an extremely excited owner shouts down the phone "It's come out, it's out, it's out!"

A word of warning though, on the legal issues. Remember that badger setts are protected by law and that it is also an offence deliberately to put a dog into a sett. No-one may interfere with a sett without first obtaining a licence from Natural England. This applies even to the emergency services, though they do not always appear to know that. In their enthusiasm to help, some have been known to start digging into a sett to rescue a dog almost as soon as they arrive on the scene.

Do not worry about the dog surviving without food or water. Hopefully, it will be out in less than 48 hours and dogs can go for days without food. They may lose a little weight – which might actually help them wriggle free – but, once they are out, they will soon recover with a little tender care.

Reproduction

Badgers mate in the spring, early summer and autumn, but they have a curious control system (delayed implantation), which ensures they rear only one litter a year. Badger cubs are born underground between December and April and usually emerge approximately 6 to 8 weeks later. The earliest I have seen cubs above ground was the 14th of February, Valentine's Day. Very soon after giving birth, the sow is receptive again, mating takes place (sometimes with more than one boar) and the eggs are fertilized. But the foetus does not begin to develop immediately. Instead, the fertilized eggs remain in the uterus unattached and do not implant until some months later. Normal pregnancy then takes about 8 weeks.

What at first sight seems a rather curious system is, in fact, a very clever trick – a natural birth manipulation system that allows the badger and other members of the weasel family to improve the survival chances of their offspring. A lot still needs to be understood about delayed implantation, but it appears to be the case that the sows are able, in some way, to determine or at least influence when implantation takes place and therefore when their cubs are born. For a wild animal, which has only one litter a year, usually of two to five cubs, that kind of control mechanism is very useful, faced as it will always be with significant fluctuations in the weather, variations in the amount of food available and as was once the case with badgers, predation by larger animals.

Delayed implantation also means of course, that if mating does take place in the spring and the boar subsequently dies through injury, disease, old age or, as used to be the case with badgers, through predation, the sow can still breed successfully. With the exception, as I have already mentioned, of foxes killing cubs, natural predation is no longer a significant factor in the life cycle of the British badger, but unhappily they still have one real enemy: man! Though shooting badgers is illegal, it does happen and so does the vile practice of badger baiting. Many badgers end their days being shot, in a baiting pit, or being clubbed to death by diggers (more on that later). Many, many more (thousands every year) are killed on our roads, although not all of the badgers found at the roadside are victims of road traffic accidents – all too often, badgers that have been shot on farmland are dropped onto the road to make it look as though they have been run over. Nature could not have intended it, but the delayed implantation system works in a way that partly minimises the impact of this modern-day wildlife carnage on our roads. It is worth noting here that there are always young, virile males waiting for the opportunity to mate.

How many times, I wonder, are the cubs we see in the spring the progeny of boars killed many months before on our roads? How often, too, are they the offspring of more than one boar? We know multiple mating happens: genetic fingerprinting has proved that beyond doubt, but how frequently is not known. Could it be that the loss of a mate, probably the dominant boar in a social group, prompts the sow to mate again? Certainly, it is clear that the sow chooses who will father her young, as is the case with all the mammals I have studied. My experience with pig breeding showed that up to three sows would prefer to mate with the same boar, even though there were as many as 16 to choose from.

As a badger watcher, there is one sure sign that I look for each year to tell me the birth of cubs is under way. From about mid-December, the pregnant females take green vegetation such as moss, grass, ivy, bluebell leaves and other similar material into the sett. If it is a multi-entrance sett, they will choose certain entrances repeatedly to take the bedding down into the sett chamber they have chosen as a nursery and it is from these same entrances, weeks later, that the cubs will first emerge.

Pregnant badgers seem to carry out similar nursery

preparations to pigs. If they are given free-range facilities, 12 hours before farrowing, sows will collect huge quantities of vegetation, grass, nettles, dock leaves and any wood or other material they can carry in their mouths. They usually finish up with enough to fill three or four barrow loads. Because it is mostly green, the material heats up within a few hours, just like any compost heap, and for the piglets and badger cubs, the result is the same: essential, life-sustaining warmth. With badgers, the business of collecting green vegetation does not stop with the birth of the cubs. It continues for weeks, often until as late as June and experience has taught me that, if it suddenly stops earlier than that, then it is likely that the cubs or the sow have been killed. One natural sett and one artificial sett that I studied over a 5-year period displayed all the signs that cubs were underground and foxes were also present in part of the sett from February to June. Both setts showed a decrease in badger activity in April: bed collecting stopped but fox activity increased and on two occasions subsequently the skulls of badger cubs were found at the entrance. The reason we often find skulls of mammals and birds is because they do not contain bone marrow, as do the rest of the bones and so they are not consumed.

Cubs seem to do everything instinctively. Once they emerge above ground, they groom themselves (though mum also joins in), they scent mark one another, practise digging quite deep holes and quickly mimic bed collecting, making use of any materials that are handy – twigs, leaves, grass, anything that is movable and some things which they find out are not! The behaviour is the same with orphaned cubs reared in captivity, even though they have no other badgers to learn from. Similarly, instinctive behaviour is evident with adult badgers. For example, if they find an obstruction on the track that they want to follow, they won't be deterred. Whether it is a fence put up by a farmer to confine livestock, a ditch or a watercourse that has been cleaned out and deepened, or a new road or some form of building development, they will keep to the same course. Sadly, that same instinct can be their downfall, of course and if a new road cuts across an old badger track, they will still want to follow it. All too often that is when they end up as road casualties. Badger-proof fencing and wildlife underpasses are now incorporated into most new road schemes.

Head over heels – but it gave me a clue

There's a lot we don't yet understand about badgers, but we are learning all the time, all of us. To illustrate the point (and to add another piece to the badger tunnelling jigsaw), let me describe two incidents, both of which led me to a conclusion that I think is new and quite important.

Years ago, I was asked to carry out a survey in connection with a proposed quarry extension to make sure that badgers would not be endangered by the work. The land had been bought some 10 to 15 years previously, the hedges were wide and unmanaged, the ground covered in gorse, brambles and long grass. Alongside was a busy quarry and there was plenty of evidence of badger activity. Night workers had become accustomed to seeing badgers and the men often threw scraps of food out to them and there were plenty of badger paths as well as roadways and vehicle tracks, both of which badgers used regularly. Nearby was the top of the quarry, a lovely wild area with plenty of signs of fox activity, but nothing to indicate badgers. This was odd because only 200 metres away was a main sett at the quarry entrance. I found two fox earths, both of them in sand, but hard though I searched I could not find a badger track, or any sign of them feeding.

It worried me, had I missed something? Surely, there had to be an outlier (a smaller sett used by badgers from the main sett nearby)? I was tired, it was the end of a long day, and so I sat down to fathom it out, and almost without realizing I took out my penknife and cut out a small square of turf to look at the root structure of the grasses. Mixed in with the roots was gravel and I started prising out each individual pebble. Suddenly, the penny dropped. I thought to myself "They can't dig a sett, pure gravel is outcropping on this entire area and it won't be suitable either as a foraging ground as there are no earthworms. I haven't found any badger signs because there aren't any, simple as that!" That brightened me up, but just as quickly I was concerned again. Close by were fox earths in what appeared to be almost pure sand. So how could that be? I went over to look at them again. There they were in the brambles, fresh, well-used fox earths. I stepped back pondering and promptly fell into a trench! Excavated some years ago as an exploratory trench by geologists, it had become overgrown, covered in weeds and brambles. It was 2.5 m deep, so I hit the bottom with quite a crash. Feeling a bit shaken, I scrambled out and then almost jumped for joy. Looking back into the trench again, I could see it was all gravel at the top, the sand had come from about 2 m down and the fox had made its earth in the sand that had been excavated from below that top layer of gravel. Enough sand had been excavated for the fox to make an earth, but there was insufficient sand for a badger to establish a sett. In terms of the survey I was carrying out, I could now rest easy. I hadn't missed the badger signs.

Now let's jump forward a year or two. Water supply companies are among the many businesses that have to take great care to ensure, as they lay down new pipelines or dig out old mains to reline or replace, that they do not destroy setts. On this occasion I was guiding a mechanical digger that was excavating the trench for a 40 cm water main to be laid at a depth of 2.5 m past an area containing a badger sett with 20 entrances. I had examined the area carefully and was satisfied that the work could be done without damaging the sett or injuring any badgers and I had taken full responsibility for the welfare of the badgers (as I always do), insisting, of course, that my instructions were followed to the letter. On the day in question, I was at the site in good time and excavation was still some 100 m away. As I had time on my hands, I was able to watch the mechanical digger taking out a cubic metre of soil at a time and having a passing interest in geology, I looked closely at the soil strata. It was the division of the second bedding plane which interested me – I saw earthworms falling from the side of the trench at the point where the two soil types, light loam and small gravel, met 1.8 m down. As we approached the sett, this division in the strata was only 1.3 m from the surface. I studied the freshly excavated soil that the badgers had deposited on their spoil heap and found that it was a mixture of the two: light loam and small gravel, so the tunnel was running into, or along, this strata junction.

That was food for thought and it has prompted me ever since to look very carefully at setts, both those which have to be dug out to allow some form of development to go ahead and setts which the police and the RSPCA have called me in to examine when badger diggers have been active. When badger diggers finish at a sett, they often backfill the hole they have dug, so it is important for the sake of the badgers in that sett to re-excavate the hole. If a tunnel has been sealed off by the returned soil, the badgers may suffocate, for as they try to dig their way out, the loose material above falls in on them. The effect is much like the action of an old-fashioned egg timer: eventually the badgers will run out of space to dispose of the falling soil and depending on the layout of the sett, may also run out of air. Invariably when I am doing this work, I find the badgers have dug their tunnel system where two different soil types meet. This happens, as you would expect, at various levels, anything from 0.3 m to more than 14 m down. Where the differing soil types meet, it is common to find extra moisture, more organic particles and looseness in the composition, which seems to provide ideal conditions for worms. The way in which these differing layers split away from each other as they are manually or mechanically separated during digging also confirms that there is a structural weakness which the badger exploits as it looks for the easiest conditions in which to tunnel.

Chapter Four

Badgers in the family

Badgers – every one a character

The more you study them, the more you realise that badgers, like people, are individuals, every one of them. Some are extremely nervous (old sows in particular), while others are as bold as brass. I well remember one little female badger that used to come up and have a darned good look at me. Often, she would scent-mark one of my shoes, just as if I were one of her social group; sometimes, she would wander off into the night, and then, without warning, return from another direction and just amble by, showing no fear whatsoever. At most setts, I make it a rule not to feed the badgers to ensure they don't become too used to me, but I did make an exception at this young sow's sett. One night I was feeding her bits of peanut butter sandwich and she'd only had a few mouthfuls when she suddenly trotted off. I thought it was a bit unusual, but within minutes she was back, followed by two young cubs. They were both very hesitant, but she trotted boldly towards me and turned as if to say "Come on, the food's over here." As she started to eat, they came forward very nervously, but then retreated out of sight, so off she went, returning again with just one in tow, but again it wouldn't eat.

This coming and going at a sett can be quite a puzzle. Looking back at some of my old notes from when I first started watching badgers, I see entries to the effect "I saw three badgers tonight, or did I see one badger three times?" It is not easy when you are new to watching. Now, more often than not, I know whether it is the same one. Markings are useful in helping the watcher to distinguish one badger from another: the width of the head stripe, a variation in colour, something different around the ears or mouth or tail. But, the way I distinguish them is more by temperament and behaviour. Some, as I say, are cautious, some are bold, others are clumsy and some seem to 'think' their way around.

One I recall with particular affection was 'Old Grandad'; he had the sort of temperament we've all seen in our elders. I remember one night arriving at a large sett. It had 33 entrances and was not simply a sett – over the years, it had served as a fox earth, rabbit warren, mink den, polecat den, woodmouse residence, rat nesting zone, bank vole home and even a refuge for feral cats. All those animals have emerged from some part of the sett over the years and have always appeared before the badgers have seen fit to come out and sample the night air. Old Granddad was, as I say, something of a character. Take the night I had arrived, in good time, at one of the eight observation points the sett provided. To my disappointment, my cattle were lying across three of the entrances I had planned to watch and they had been busily rubbing their heads and shoulders in the entrances and over the spoil heaps. "That's beggared my night up," I thought. However, as the day ebbed away and the light faded, the cattle moved lazily away and I decided to wait longer, hoping for that first sight of a black and white head, nose upright, sniffing the air cautiously – that was Old Granddad's routine. He would then groom and scent-mark everything in sight. Not this night! Out he came, and off he went down a well-used path, head bowed, grunting and grumbling to himself, just like an old man going for his nightly pint. He looked, for all the world, thoroughly disgruntled, annoyed because the cattle had made him late for his favourite snack.

Another badger I came to know quite well had what you might call a stubborn streak. He'd emerge at night, sniff the air, catch my scent and then pause with his head in the entrance. He wouldn't come out and he wouldn't go back in. He simply stayed there until I moved off. It was as though there was a psychological battle between the two of us, him saying "I'm not coming out until you go" and me saying "I'm not going until you come out." Guess who won. You're right, he did.

Hiya, captivating Hiya

Strange, isn't it, how chance or fate plays a hand in our lives. One day in the early 1980s as I was heading to Oswestry to pick up a tractor part, I switched on the radio and tuned in to our local radio station. As luck would have it, they were interviewing a local RSPCA inspector and the subject was badger baiting. He was explaining how diggers went about locating badgers underground by fixing tiny radio transmitters on to their dogs, which they then put into setts. The dogs crawl through the tunnels and chambers until they come face to face with badgers. There they stay, barking and yapping (often getting badly bitten) and that provides the men up above with a 'fix', a spot they can dig down to reach the cornered badger, which they then drag out, holding it by the tail, which prevents the badger from biting. Oddly, despite my years of badger watching, I knew next to nothing at the time about badger baiting, an extremely cruel sport that I assumed had gone the way of most other barbaric activities, half a century ago. There had been none in my study area, so I was surprised and shocked to hear that it was still practised in parts of my own county and more than that, I was extremely angry with the RSPCA inspector for spelling it out and describing it in such graphic detail.

My first thought was that the description would be heard and noted by those who knew all about foxing and rabbiting, but probably hadn't given a thought to badger baiting. Now they had been given a textbook description of how it was done. I was none too pleased. That evening I rang the RSPCA inspector and gave him a piece of my mind. He listened while I sounded off, then he told me about the problems the RSPCA and police were faced with. Put simply, they lacked expertise. Six times they had taken badger diggers to court and six times they had failed to get a conviction. As prosecutors, they simply did not have anyone who knew enough to challenge what the defence were putting forward in mitigation. I didn't know it then, but some years later the role of expert witness for the Crown Prosecution Service was one that I would be called on to fill from time to time. But that's to digress.

This particular conversation was to spur me into taking on something quite different. I listened as the RSPCA inspector, who later was to become a good friend, told me of another problem. There was nowhere local they could take injured wildlife to recover from injuries once they had been seen by a vet and no-one locally who had the know-how to look after orphaned cubs. To make his point, he went on to tell me about a young fox cub that needed caring for and asked if there was any chance I could help? I said I thought I could and that was the start of a commitment that was to last over many years for myself, my wife Cris and my children Tristam and Pamela.

Our home and our farm became both a home and a recovery ward for a long succession of sick, injured and orphaned birds and mammals. They needed warmth, care, affection, food, water, security and motivation. I like to think that we gave them all that and more importantly, the chance for them to be free again as soon as they were fit and ready. The most important thing about the rehabilitation of wild animals is knowing when to return them to the wild. All too often they are kept too long in captivity and that saddens me.

The first of these animals was that young fox, a fox whose photograph was to win the hearts of all sorts of audiences, even hardened prisoners. It was a fox with an unusual background and one that was to acquire an unusual name. The RSPCA had taken her into care after a gamekeeper had rung them up. He had been on his rounds of the estate that morning, a vixen had come through a hedge and he'd shot her. But then her cub appeared. "I hadn't the heart to kill it," he told the inspector and he went on to admit that the little cub had made such an impact on him that he would never feel quite the same again about shooting foxes. Quite a confession that!

Anyway, that's how this lovely little fox cub ended up with us and the next thing was to name her. My children were young at that time and they had a go with suggestions. For two weeks this went on and still we couldn't agree, and then we resolved it almost by accident. By this time, something of a feeding ritual had started: the cub was kept in a pen and fed at the same time every day and, as I approached it to drop some food in, I always talked in a soft voice to let it know that I was approaching and I would urge anyone in the same circumstances to do the same. Never approach a wild animal quietly while it's lying injured or in rehabilitation; always talk, or whistle or do something when you are some distance away to let it know you're approaching. The last thing you want to do is frighten it, or as we say in Shropshire 'fritten' it. I got into the habit of greeting her the same way each time. All I said was "Hiya", and that eventually is what we named her.

Hiya stayed with us from early April until July, growing from a tiny cub to a three-quarters-grown adult. Her stay was brief and deliberately so, for I wanted to return her to the wild. The longer she stayed with us the more likely she was to become dependent on us. I'll tell you about her release in a moment, but first let me give you just a flavour of the impact she made on others, not just in the course of a few weeks, but over months and years, long after she was released to the wild. I took a photograph of Hiya and used it in my talks to groups and audiences all round the country – WIs, adult education classes, wildlife groups, even farmers' wives, they all took to her. This little fox even won the hearts of prisoners in Shrewsbury Prison.

I had been asked by HM Prison Service if I would give a talk about badgers, but I chose not to. Instead, I gave one of my other talks called 'Return to the Wild' – ironic, in the circumstances. Arriving at 9 am I was ushered in through the gates, taken across the courtyard into a large building and up the steps to one of the top floors. As we reached the room, I remember looking out through one of the barred windows and thinking that there were eight locked doors between myself and freedom – that made me feel very claustrophobic. I was there to give a talk, yet for the first time in my life I felt trapped and caged; it was the same feeling any animal would experience.

Gradually, the room began to fill as the wardens brought in prisoners three or four at a time and with the numbers at around 24–25, I was asked if I had everything I needed. I said "Yes," and with that the wardens made to move off, announcing they would be back at 11.30 am – that was almost two and a half hours away and I had prepared a talk of about an hour! Not a good start and it was to get worse. Plainly, the men weren't interested. They sat in small huddled groups, mostly with their backs turned to me and my introductory remarks were barely audible above the hubbub; they were laughing and joking, and only a couple showed any interest at all. I had, by then, put on three or four slides and panic was beginning to set in. Fat chance of lots of questions I thought, for that's how I had hoped to involve these men in my presentation, so as to spin out the time.

Then I showed the slide of Hiya, a close-up taken of her head as she peeped through her pen. One or two prisoners nudged the others to look round. Suddenly, there was silence, and from that moment on they were captives of my pictures and my images of wildlife. I finished my talk and asked if there were any questions. They came thick and fast, and they were intelligent questions too. It was one of the best discussions I have ever experienced after a talk, with the exception of talks to 7–10-year olds in schools who always ask the most soul-searching questions. Almost before I knew it, the wardens were back in the room and the first thing they said to me after they had stopped, listened and looked round for a while was "How on earth did you do that?" Meaning, of course, how did you get that lot so interested? "Simple", I said, "I owe it all to one little fox." And I did. Hiya captivated this audience as she did so many others. Why? She was simply so beautiful. It was all there in the eye: the tenderness, the gentleness, a look to touch the heart, a look that even now draws a gasp of pleasure from everyone who sees that picture. My parting comment to the prisoners was, "If you've enjoyed the talk, as I hope you have, don't let it stop there. Go to the library; choose a book on any subject to do with wildlife. It will give you a new outlook on life and you'll never see the inside of this place again."

Let me tell you one more story about Hiya. Remember, my background is farming. Stock are reared, cared for and then sent to slaughter; that's how it is on farms. I had made my living that way for years, as had my father and grandfather before me. So, you might expect the loss, the departure, of an orphan fox to be just another event. Well, it wasn't quite like that. In the wild, female cubs often stay with the vixen. Male cubs, by contrast, start to fend for themselves by late summer or early autumn. When the end of July arrived, I thought it was time to release Hiya to the wild. She deserved freedom, she would have to learn to fend for herself, but she would do that instinctively. We had quite a few foxes already on our farm and I didn't want to release her there, but my brother, a dairy farmer, lived only three miles away and didn't have many foxes on his land, so I decided to try her there. I popped her in a box and took my camera, determined to record the moment of freedom: it was the best photograph I never took.

Hiya was quite a nervous animal and I had never handled

her, so when I opened the lid I expected her to be gone like a bolt from the blue. But no, she stopped in the box for a few minutes and that gave me time to walk away, camera at the ready. She came out and looked towards me and that was it, I couldn't keep the camera steady, I was simply too emotional. She looked at me as much as to say, "Where am I, what do I do?" She took another couple of paces, stopped, turned her head towards me and this time the expression was a mixture of both delight and disbelief. "Am I free?" she seemed to be saying. She then took another four or five strides, stopped once more, looked at me again and cocked her head to one side as if to say, "Thanks". Once more, she trotted on and again she looked round. Hiya was a full 50 m away before I took my first shot of her. On she went, through the middle of a herd of cows, more than 150 of them and not one looked up. 200 m away she stopped, turned and began to come towards me and that's when the stockman in me took over. I clapped my hands, shouted and, with just another glance, she turned and headed off again. Much as I regretted it, she had to go. I couldn't let her come back. It is not easy to let go, but you have to.

I was emotional then and I still become emotional now, thinking of her, our first waif and stray. Others were to follow, of course: more fox cubs, badgers old and young, polecats, stoats, weasels, owls, kestrels, sparrowhawks and buzzards. But Hiya will always have a special place in our memories. So, too, will the badger that our children christened 'Bodger'. There were lots of other badgers, but only one Bodger. More of him later, but let me just briefly mention the little owls, perhaps the luckiest to survive of all our visitors.

Shortly after Hiya arrived, the RSPCA inspector rang me up and said, "George, I've got two little owls. They were born in a barn in a farm baler and were covered in grease when they were found. We've cleaned them up and they are not yet ready to fly. Can you look after them?" He brought them to the farm along with a supply of dead day-old chicks, which are often used as food for birds of prey in captivity. But, I thought there had to be another way, a more nutritious food, so I picked up a road-casualty rabbit one day. There's enough here, I thought, to last them for weeks. So I'll cut it up into steaks, fur, flesh, blood and guts, the lot, and put the bits into bags, pop them into the freezer and take them out as and when needed. Fine, good idea, I thought. Like an idiot, I put this rabbit on the chopping block and got an axe to it. Within seconds, of course I was covered in blood, bits of fur and other unmentionables. I stopped and thought, and gave myself a proverbial kick up the backside. What a fool I'd been. What I should have done, of course, was to put the whole rabbit in the freezer, wait for it to freeze (that, as I found out, took about 48 hours) and then set about putting it in a vice and sawing it into steaks. That proved to be clean and easy, of course and raw rabbit was much better nutritionally. As a further bonus, it wasn't long before I had my wife fully trained, scraping up similar road casualties! Between us, we did pretty well collecting this free food, though I was apt to blot my copybook from time to time. Occasionally, I'd pick up a rabbit, put it in the boot and forget all about it. Do that in the middle of July and nature has a way of reminding you a day or two later with a smell that lingers long after the offending carcass has been buried. Those rabbits served us well over the years, not just for the owls, but also for other birds of prey, including a number of kestrels.

One kestrel came to us, again through the RSPCA, still recovering from severe shotgun injuries to one of her wings. They had patched it up with a splint, but when she came to us she could only walk; she couldn't fly. I questioned whether we should have been rehabilitating her at all, but as the weeks went by she grew in strength and began to fly, at first only a few metres, but gradually further and further. I put little posts at various heights inside the pen that she could perch on, but it still took several months before she could fly the full length of the pen, which was 12 m long and 3 m wide. Winter came and she wasn't ready for release, so I kept her until the following spring. By then, she was much stronger, but still her one wing dipped and I was anxious about her prospects. Kestrels hunt by hovering and I wondered if this one would be able to cope. There was only one way to find out, so I opened the top of the pen to allow her to go. For three or four days she did nothing, then one morning the pen was empty and she was nowhere to be seen. I put some raw rabbit on the top of the pen in case she returned and needed food, but I didn't see her for another three months or so. A visitor saw her perched on the top of the pen and I recognised her immediately, for that left wing still dipped a little. Weeks later, I saw her again and from time to time over the next couple of years she was seen hunting. Since then, kestrels have made quite a recovery in our area and I often wonder whether some of the youngsters I've seen are hers. I hope they are.

Two other observations, gleaned from our days as rehabilitators, are worth recording. I have always found that the females of any species adapt to their captive environment much more readily than males. With kestrels and sparrowhawks, for example, the males are more reluctant to feed and sometimes we had to force-feed them to get them started. There was also something about behaviour, both in birds and mammals, which told me from time to time that I was dealing with a creature that had been top of its hierarchy – this was especially noticeable with foxes and badgers. During the summer, the smell of a fox is very weak, irrespective of whether it is a dominant male or not and often you could walk past pens containing two or three foxes and pick up no scent at all. In the autumn it was so different and even visitors to the farm would notice. They would sniff the air and say, "George, you've got foxes here."

As I've said, my sense of smell is very good, and at times I could pick up the foxy smell 50 m away, and almost always it originated from a dominant animal, a superior example of the species. Invariably there was something special about the look, the way the animal held its head that told me we were caring for a dominant member of a social group. It was the same with the vixens. With some, you wouldn't know from the smell that they were foxes, while with others the odour was strong, pervasive, and they also recovered more quickly. Again, there was something different about the way they moved, their posture. It was a bit like being confronted by the catwalk models of the fox world or by a world champion boxer. The same was true of badgers: dominant boars had a particularly strong body odour and stood proud. They were much more positive in their movements, their eyes would follow your every movement, and there was always an air of mistrust. Dominant sows would be ever poised for that split-second open-mouthed lunge towards you and their recovery was much quicker.

Bodger, the one and only Bodger

Early one spring morning, I received a call from a gentleman in west Shropshire. Whilst walking his dog, he had come across what he believed to be a badger cub huddled under a hedge. It was a bitterly cold morning at the end of March, so I lost no time in heading off and eventually found my way down the network of tiny lanes to the badger. I needn't have bothered with my carrying box; this little lost soul was so tiny – barely alive. Without medical help he had only hours to live. I thanked the caller, refused his offer of a cup of tea, picked the cub up, wrapped him in my sweater and drove off to the vet with the car heater full on. The vet gave him a couple of injections and told me to keep the cub warm and hope for the best.

Arriving home, I found we had visitors – two ladies we had got to know through our involvement with badgers. As Cris set about making us all a cup of coffee, we talked and as we chatted I noticed the young cub was infested with fleas, so I started to dust it with flea powder. The coffee arrived and there I was, cup in one hand and flea powder in the other. I glanced up to be greeted with disproving glances from both our visitors. Tongue in cheek,

I asked if they had ever tasted flea powder? "No", they said, very emphatically. "You've missed nowt," I told them with a smile, returning to the task in hand. Soon we had our young arrival tucked up nice and warm, though I was none too hopeful that he would even make it through the night. But he did, and his energy the following day was astonishing – overnight, the omens had turned from bleak to promising.

In no time at all, Bodger, as Pamela and Tristam decided to call him, was a member of the family. In those first few important weeks of recovery he stayed indoors for warmth, as did all our young boarders; young animals must not be kept in solitary confinement. At night, Bodger loved to be groomed. He would lie on my lap while I combed him round the ears, underneath his legs and belly, then, when he had had enough, he would simply lower himself gently backwards off my lap and scurry under the settee. He would remain there, unusually quiet and it was quite a few days before I decided to investigate. One night, I pushed the settee to one side, only to find to my horror that he had been mimicking bed collecting; he had lifted the edge of the carpet, pulled bits off and gathered them into a heap. Not exactly behaviour to endear him to my wife, but very interesting all the same. Bodger was far too young when I rescued him to have learned bed collecting from his mum, so his behaviour had to be instinctive. What is more, it was happening at about 8 to 9 weeks old – mirroring cub activity that I had seen many times in the wild. To me, that suggests that, for animals with a relatively short lifespan, behaviour, for the most part, is not picked up from the parents. Much of what happened in our house in those formative weeks mimicked behaviour in the wild.

At the time we took in Bodger, we were also looking after some baby fox cubs and with no adult badgers to intrude, all the cubs would romp and play. The fox cubs also took great delight in stalking young Bodger, pouncing on him as he browsed around looking for titbits. Cris wouldn't have approved, but sometimes

I hid bits of worm for Bodger to find and it was an object lesson to me to watch how easily he'd pick up the smell, often from several feet away – a sniff, a scurry and the worm was gone. I put a stop, incidentally, to his carpet crunching by leaving an old shirt or pullover for him to tear up. That did the trick.

Over the years, I have raised various combinations of fox and badger cubs, and, as I mentioned earlier, I have noticed that, from the age of 10–12 weeks, by which time they would be in pens outside, they become indifferent not only to each other but to humans and domestic pets. It's that teenage independence thing we all know about. They don't fight, they just ignore one another and that's when I start to prepare them for eventual release. First, they are separated out, each into its own pen and from then on no-one other than myself is allowed to see them. It is at this stage that the person who feeds them becomes, in their eyes, head of their family. It is imprinting of a kind – though not the sort that turns a wild, fiercely independent animal into something supine, semi-domesticated and totally unsuitable to be released back to the wild.

Around July, I like to release any youngsters in our care. The year we had Bodger I was contacted by members of the South Yorkshire Badger Group who were looking for badgers they could introduce to parts of their area that had suffered dreadfully from persecution. Bodger had been raised with an erythristic male called Ginger, and Lucy, a young sow who arrived a week or so after Bodger. In their early days with us, when they were all allowed to play together, the sow and the erythristic seemed to get on especially well, so I decided both could go to Yorkshire. I would keep Bodger, because it wouldn't be wise to release a sow and two boars together. For all sorts of reasons, but especially because of the (as yet unsubstantiated) possibility of transferring TB to cattle, it is absolutely vital that the correct procedures are followed in all instances involving the transfer of badgers from one area to another and today I would advise against translocation. In this instance, the appropriate MAFF (now DEFRA) veterinary officer in Yorkshire was approached and he indicated that he had no objection to the proposed transfer, provided our local vet signed what, in effect, was a certificate of a clean bill of health. That was duly obtained, and the transfer of the two badgers went ahead as planned. There were no problems and 2 years later we received the good news that three cubs (none of them erythristic) had been born at the artificial sett provided for the two youngsters by the South Yorkshire Badger Group. The sett was covered by a mesh of reinforced steel to deter any would-be badger diggers and the South Yorkshire Badger Group also took great care to keep the sett under constant surveillance to further deter any wrongdoing.

So much, then, for Lucy and her young companion. But, what could I do with Bodger? He was fit and ought to be released. But where? There were no ready-made empty badger territories in Shropshire that I knew of, so I decided to repeat the procedure that had worked with a road casualty badger until the donkey had intervened! (see Chapter 5). Bodger would be given his chance to set up home close to the farm. Hopefully, he would be accepted, or at least tolerated, by other badgers in the vicinity. Having made the decision, I left the pen door open. The next morning I looked in and Bodger was curled up fast asleep, but there were signs of foraging close by, which suggested he had ventured out. A couple of weeks went by and each morning there was Bodger fast asleep in his pen. Then it all changed. As I approached the pen one morning, I saw a huge pile of soil at the end of the pen and who should come to the entrance looking very pleased? Yes, it was Bodger. He stalked around me, scent-marking my feet as usual and disappeared back into his newly dug sett.

Soon, he was back out and off he wandered into the garden, looking for the titbits that we often left for him and as he ambled away, I realised now was my chance to find out how much he had dug overnight. So I hurried off, collected my drainage rods, one of them complete with a small end wheel to ease it through drains and round corners, and pushed the first rod down. Then

I added another and another and another, until eventually the rods reached the end of the tunnel. I pulled them out and measured them – 5 m! In one night Bodger had dug 5 m of tunnel. Just imagine what a family of badgers could achieve in a few nights. Well, with that little lesson in tunnelling expertise well and truly learned, we kept a close eye on Bodger. Next morning, he was back in his own pen, but within a week he had gone native, so to speak, preferring his own natural sett to live in, no doubt by then complete with at least one chamber.

We had a visit from a wildlife police officer from the Isle of Wight who stayed with us for a few days. He was keen to increase his knowledge of badgers, so I took him down to see Bodger. Normally, I would call his name and Bodger would come out quite cautiously, sniff the air, as badgers do in the wild and then he would amble towards me and scent-mark my boots. Not tonight! Bodger came out without pausing at the entrance and then charged at my legs, biting ferociously, I had wellies on, and he began tearing these apart. There was nothing else for it, we had to get out of there double quick, so we simply ran for it and he was still on my heels as we got to the drive. But, as Bodger reached the drive, he stopped, paused, grunted, then turned and headed back to the sett. I was absolutely stunned by the incident; it was so out of character, so unlike Bodger. I waited an hour and then I headed back to the sett, armed this time with a broom, not to hit him with, simply to use as an extension of myself. If he wanted to bite something, he could bite the broom. I went down and called, and out he came again, taking mouthfuls out of the broom in a split second! I dropped the broom and ran, and again Bodger came after me, and again he stopped at the edge of the drive. That was enough for one night; I decided to give him time to cool down. Next evening, armed once again with the broom, I headed towards the sett and some way off I called to him. Out he came, and once again he charged at me. Once again, I dropped the broom and headed for the house and he followed, but again he stopped at the drive. That's when the penny dropped;

clearly in his mind the drive was the boundary. Much of the garden, together with the pen and the sett was his, the drive and what lay beyond was mine. I guessed, too, that Bodger had female company in the sett. More than that, he was lord of the manor (as I had once been to him) and was intent on proving it. I could think of no other rational explanation. Perhaps unknown to me, he was also involved in some skirmishes with another boar threatening his dominance over the sows. One thing was certain, however: from his days as a young cub, he had looked on me as top badger, but now he was challenging my supremacy. Whatever the explanation, Bodger had changed. He would still come and visit, and still ask for food, but his pen and his sett were now part of *his* patch.

Every evening, Bodger would come up to the house at dusk, though never if we had visitors and we'd throw him a titbit – a biscuit, a leftover scrap of food – and almost before we knew it, this became a routine. Then, around 3 am one morning, I was woken by a knock on the door. Cris heard it, too. Whoever could it be at this time of the morning? We listened and there it was again, another knock, followed by three short sharp raps and a pause, then more knocks. I put my dressing gown on and went down rather hesitantly. "Who's there?" I called. No reply. I opened the door very gingerly and through the gap at the bottom of the door popped Bodger's snout. I then did something absolutely stupid – I threw him a biscuit! He grabbed it and wandered off. I shut the door and stumbled half asleep back upstairs.

Next evening, around his usual time, Bodger was there for his titbit. Fine. But about half past two in the morning what happened? We were woken again by a noise at the door, it was Bodger rattling the letterbox. This time I stayed put. He'll go away, I said to myself. But no, Bodger was nothing if not persistent, he kept rattling and banging, and eventually, again stupidly, I gave in, wandered down, threw him a biscuit and off he went, looking a bit smug, you might say. Round two to Bodger. But, at least I was learning.

Next night: rattle, rattle. This time, I really did stay put and eventually Bodger gave up, grunting and grumbling to himself as he ambled off. The next night: same scenario. Again, I stayed put. This went on for about a fortnight, Bodger rattling away at the letterbox and the two of us sitting tight, pulling the bedclothes over our ears to drown out the noise. Eventually, Bodger got the message and the middle-of-the-night visits stopped, but still he came for his late evening titbits. Then one night, really without thinking, just as I was going to bed, I opened the door and threw a bit of uneaten apple pie out on to the path. In the middle of the night, what happened? You've guessed it: rattle, rattle! But, this time, the noise was somehow more persistent and it sounded louder. "Oh, hell," I thought, "It's Bodger again."

But, as I listened, I began to have doubts, it could be someone at the door, and I had better go and see. Down I went, only to see Bodger's paw reaching through the letterbox, moving this way and that, obviously searching for something. I stopped and watched and then I spotted the cause of all this commotion. When I had thrown out that bit of uneaten pie, some of the crust had fallen on to the carpet on our side of the door. Bodger could smell it, wanted it and wasn't going to be put off by something as minor as a door. As luck would have it, my camera was on the hall table, I grabbed it and quickly took a shot of this paw reaching through the letterbox. Then I stopped and waited, and again the paw began to search so I took more shots. Eventually Bodger's persistence paid off, the paw found the crust, grabbed it much as those miniature, remote-controlled grabs do in the kiddies' amusement arcades and triumphantly lifted it out. That achieved, Bodger wandered off. Again I'd learned something: Bodger was by no means stupid.

Well, after that, we drifted back into the same old routine of late evening visits, but thankfully we were spared the early-hours 'knock-knock, who's there?' pantomime. Until, that is, one summer's night about two years after Bodger was first given his freedom to roam. Again, it was about 3 am, when we were both woken by snarling and scuffling, and the sound of bodies thrashing on the loose gravel of our drive. I stumbled, bleary-eyed, to the bedroom window and heard enough to confirm that it was badgers fighting. So I hurried downstairs, grabbed my wellies and a torch and stepped out into the night air, clad only in my pyjama top. As the beam of the torch cut through the gloom, I saw two badgers locked in combat. I moved towards them and the intruder (big enough to be another boar) fled into the night. I half expected Bodger to give chase, but instead he ran to me and began to criss-cross my feet in a state of great agitation. He was covered in blood and dripping white with sweat. I could see his ear was badly injured and so too were his jaw and his lip. I thought it was too risky to pick him up, so I began to walk slowly towards the pen. Instead of walking with me, Bodger half threw himself across my feet and I found myself shuffling all the way, rather like a penguin trudges through the snow, my feet weighed down by Bodger, who was all the time turning and scent-marking my wellies – it was quite extraordinary behaviour. Eventually, we reached the pen and Bodger immediately buried himself in the straw. There was nothing more I could do for him at that moment; I felt sorry for him, but wounds apart, he was fit enough.

Next morning, I looked in on him and he was sound asleep. His ear was split and he had a very nasty gash across his jaw. I wasn't worried about the ear, but I wasn't so sure about the jaw. So I rang the vet, described what happened and his first reaction was, "You'd better bring him in." But I hesitated. I thought to myself, this sort of thing happens frequently in the wild, badgers are often injured fighting and yet they recover fast enough. So, I said to the vet, "Look, I can see Bodger at least twice a day. If the wound doesn't heal and starts to go wrong, can I bring him in to you? Can we leave it to see what happens?" So, that's what we agreed to do. The following morning, I looked in on Bodger and would you believe it, the wounds looked as if they were a week old, they had mended so well. It was really quite startling. In about

4 or 5 days, all that was left was a scar. If I hadn't seen it for myself, I wouldn't have thought it possible. So, that was yet another lesson, thanks to Bodger. Wild animals do have this remarkable capacity, provided they are fit and healthy and eating the right food, to recover from injuries that, to us, look awesome.

It also set me thinking about other badger injuries I'd seen: badgers with septic wounds that wouldn't heal readily. Those animals frequently needed antibiotics and other forms of care from veterinary surgeons to get them on the road to recovery. Why? It didn't need much working out, Bodger's were first-time injuries; the other wounds I'd seen on much more seriously injured badgers were new wounds on top of old wounds, they were injuries inflicted not over one or two nights, but at intervals over many months. Resilient though the badger's body is, there is a limit even to its powers of recovery, if subjected to repeated punishment. One conclusion led to another. As I've said, I had never really subscribed to the view that the bulk of the injuries we see on badgers are invariably the result of territorial disputes – that is, a badger from one social group fighting another from a neighbouring social group to lay claim to a disputed piece of ground. A territorial skirmish would often be just that in my view, a one-off event, probably quite limited, with the defeated badger retreating to lick its wounds. Repeated fighting above ground by 'opposing' badger social groups would also be well documented if it happened that regularly. I am sure that, most often, the answer to badger injuries lies elsewhere. The wounds that don't heal, the bodies that bear witness to frequent fighting, point to quite another explanation in my view. They are badgers that, too often, have been locked in combat underground, the battles have not been territorial but hierarchical, part of the frequent struggle in the animal kingdom to determine who is 'top dog' or, more accurately in this instance, 'top boar' or sow in the social group.

The nature and location of badger wounds provide important clues. They fall into two groups: wounds to the head and neck, and wounds (often very severe) to the top of the rump. Wounds to other parts of the body are comparatively rare. Badgers meeting head-on inside a tunnel would inflict wounds to the head and neck, while a badger being pursued by another inside a tunnel would be bitten on top of the rump. I am convinced that this is what happens more often that not and it explains why we so frequently find badgers with old wounds and fresh wounds, all more or less on the same place. A badger harassed in this way, subjected to repeated fighting, would clearly suffer stress as it slipped lower and lower down the social scale. Stress lowers the body's natural resistance, reduces its ability to mend wounds quickly, then septicaemia sets in, bringing with it, in severe cases, a form of delirium. I know it happens, I have seen it too many times.

I have been called out to badgers in distress in some unlikely locations. One was a bus station in the middle of a town, where the badger was badly wounded and plainly disorientated. On another occasion, I was called to Oswestry Police Station; a badger suffering badly from fighting injuries had walked into their dog compound. In another instance, a local farmer rang me to say they had caught wind of a disgusting smell coming, would you believe it, from an outside loo. Apparently, the loo had been disused for more than 30 years and, when they cautiously investigated, they found a badly wounded badger cowering in it. Another badly injured badger took refuge in a greenhouse and yet another was found in distress, sheltering in the front porch of a private house. Farm outbuildings and coal bunkers are two other places where they often take refuge. All the badgers I have just described shared the same type of symptoms – delirium brought about by intense pain and severe injuries sustained over many months or even years. Cris and I have rehabilitated these animals, which more often than not were old, with teeth worn or missing, claws lost or damaged and bodies severely scarred. They were sent to us after the vet had tended their wounds, repaired broken limbs and stitched badly torn ears and chunks of skin. We did all we could for them, but I have often asked myself, is it right to try to save them? They are badgers that have lost out, animals that will always be second best. What happens to them when we return them to the wild? Do they have a life worth living? Or, do they simply suffer again and again? Would it be kinder therefore, to put them down? I feel we should help those animals which have a realistic chance of being returned to live independent lives in the wild.

But, back to Bodger, who by now was big enough and old enough to be regarded as a threat to the local hierarchy. The injuries received in his early-morning scrap on our drive healed unbelievably quickly, so I thought, right, Bodger, you're not stopping in this pen for the rest of your life, you are going to get back out there in the big wide world. I had to be strong to be kind. After about a week, I left the pen gate open and walked away, thinking that, if the worst came to the worst, Bodger, if he was beaten up again, would simply run home to me. So, I wasn't too worried about him. Fortunately, there were no immediate further similar incidents. Occasionally, he would come back to the door and wait for a titbit or two, but one evening, instead of eating it, he picked it up and hurried off, head held high and disappeared inside the sett he had dug by the pen, a sett that by now boasted a large spoil heap containing the equivalent of 15 barrow loads of soil. Within minutes he reappeared, trotted up to the door and rattled the letterbox, plainly asking for food. I threw him some more and off he went back to the sett. Next day, with the light almost gone, Bodger came a-knocking again and this time I threw him a larger chunk of bread, grabbed my torch and shone it on him as he headed, not to his sett, but off across the garden into the field in the direction of the main sett and the haunt of his old adversary some 200 m away. Intrigued, I managed to keep track of him and suddenly, before he reached the big sett, the beam of the torch picked up two more black and white heads. Bodger had at least two companions. Were they young immature badgers or older sows? I couldn't be sure. And, where was the old boar?

I decided to keep a watch on the sett. A couple of nights went by without a sign of life – nothing unusual in that, it often happens. Then, on the third night, a head popped out. It was Bodger's, and in no time he was joined by two others, both mature sows. So it *had* happened. Bodger's big day had arrived, the main sett was his now. The two sows with him were, I felt certain, the two that had been accompanying him to our garden sett and that would explain, in part, the huge spoil heap. What, I wondered, had happened to the old boar? We'd had one or two casualties on a major road about a mile away, so perhaps the old boar had ended his days there. Perhaps he simply died of old age or had been shot. We can only guess. Whatever the explanation, Bodger's time had come, he was now top boar and he had achieved the lifestyle we thought would pass him by.

All went well for another couple of years. The sett in the garden and the main sett in the field both grew in size, with more entrances, and more spoil. Bodger and his companions were using both, we felt sure. It was all too good to last. One night, in early August, Bodger came for his usual crust of bread and then disappeared off through the garden; by now it was pitch dark. A little past midnight I heard a single gunshot ring out not 100 m away. I didn't take too much notice, for locals quite often go out lamping after foxes and rabbits at night, using a very powerful lamp to pinpoint their quarry, which is then shot with a rifle. I have not seen Bodger since. I can't prove it, I can only surmise. But, I believe Bodger died that night at the hands of lampers. For some time, there had been rumours that two or three well-known locals had been shooting badgers and it was said they had been overheard bragging in a local public house that they had killed seven badgers in one night. There were tales, too, of strangers from London, who would arrive in a van at the weekend, stay for a day or two and then drive off. Several times they were seen in the company of these local villains and more than once they were seen loading sacks into their van. What was in those sacks? Could it have been badgers, Bodger among them? Again, it's only rumour, but there have been claims made of an illegal black-market trade, centred in London, in badger parts, sold on for alleged medicinal properties. I don't know the truth of it. All I do know, as I say, is that the night of the single shot was the last night I saw Bodger. I know, too, that in a very short time we had no badgers at all; there were no more badger road casualties and badger paths that had once been worn smooth from constant use now began to grow over. One large local active main sett, which I had studied for years, became disused and overgrown, for the first time in my memory. Everything had changed and suddenly too. I reported my suspicions to the RSPCA and the police, but with resources stretched they had neither the time nor the money to carry out the surveillance necessary.

Bodger was truly one of a kind and although he's gone, he'll never be forgotten, as memories of him live on, and so do the videos of his television appearances. As a cub, he was featured many times on nation-wide television.

Chapter Five

Badger rescue

Road victim with a sad secret

Animals tell us such a lot from their behaviour, especially from their movements and their facial expressions. Badgers aren't as visually expressive as foxes or other animals, but the signs are there, as I learned both by watching activity near setts and during our rehabilitation work.

Over many years, Cris, the children and I have cared for and released well over 100 badgers. Every animal was different. One in particular was a road casualty; she came to me from a vet and was brought in with concussion. The vet thought that, in a few days, she ought to be ready for release, but I soon saw that this badger was different. She walked around the pen very slowly and hesitantly, she would eat and drink, but I knew something was wrong, so I decided to try something. I moved close to the pen and clapped my hands. This frightened the badger and she ran towards the sleeping compartment, but she missed by about 30 cm and went headfirst into the side of the fence. I realised then that badger could not see, so I took her back to the vet, who examined her and he agreed – she was totally blind. I suggested to the vet we might as well put her to sleep. "No, George", he said. "It might have been blind before. It might have been blinded in the accident. We don't know. Give it some time. Give it six weeks." So we did. Sadly, there was no improvement in her sight, although she did begin to get around the pen more briskly, so we gave her another month, but again there was no noticeable change, so it was decision time. Again, I thought we ought to put her down. The vet thought differently. "Look, George," he said, "At least it's alive – it's living a good life. It's used to its pen." So we kept her and she lived about another 18 months before dying of old age. I learned a lot from that old badger.

Did we do the right thing? Should we have released her? Do you put a blind badger back into the wild? If we had been certain that she was blind before the accident, then I'm sure that that's what we ought to have done once she had recovered, for there is no doubt in my mind that, blindness apart, she was in all other respects fit enough physically to be released within a few weeks of the accident. But, of course, we didn't know how long she had been blind. In retrospect and with the benefit of years' more experience, I think we should have put her back. Badgers don't need their eyesight to survive; they can get by just by relying on their sense of smell.

As I say, I learned a lot from that badger, especially about emergence times. In the winter, the wild badgers were unpredictable: you could never be sure if or when they would come out. In December, our blind badger would invariably come out about half an hour before dark to urinate and drink, but as the days lengthened she came out later. By June and July, and I had her for two summer seasons, she would not come out until 11 pm. She couldn't see, she couldn't distinguish light from dark, the vet was sure about that, so what was it that told her when to come out? We'd call it the body clock, but how does this work? What is it that switches on that body clock? Is it changing temperatures? Is it the pull of the moon? I'd very much like the answer to that one.

When you spend years caring for badgers, other characteristics become apparent – some are very fussy eaters, some foods they will eat, others they won't. Some will settle down comfortably, lie with their paws forward and take their time, while others will bolt down their food and empty their dish almost as soon as it's put down. We tried them with most of the foods we humans eat: tomatoes, cucumbers, onions, bananas and lettuce they did not like. Cooked potato, carrots, sprouts, peas and gravy they would gorge on as if there were no tomorrow.

The more you work with animals, the more you learn. For example, animals will always give a warning signal before they flee or attack; with pigs, the boars will scrape the ground with their front feet, open and close their mouth rapidly, sharpening their tusks, frothing as they do so and make a barking sound caused by a large expulsion of breath. As long as they continue to do this, you are comparatively safe, but even so it is wise to take the hint and leave. If you choose to stay and the boar stops what it is doing and freezes like a statue, looking at you with a fixed stare, then take it from me, it's a split second away from attack. Use that split second to get the hell out of there!

With badgers in captivity, there are also characteristic warning signs that something is going to happen. If, as you get close to a boar badger, it holds its head very stiff and turns towards you, that's a sign it is about to attack. If it turns its head slightly away, you know that, in all probability, it is going to run away. With females, you get no warning at all; their reflexes are much faster and they will snap as quick as lightning. I have seen much the same sort of behaviour with badgers in the wild. Over the years, it has fallen to me to rescue 18 badgers trapped in snares. Given the boar's proven reputation for fighting other boars, you might expect that a snared boar would be more dangerous than a snared sow. In fact, the male is the easier to handle, as for some reason males don't show the same degree of fight, fear and anger. Females fight fiercely and will snap viciously, again without warning. I well remember one young lactating sow cruelly snared. I was called in to help, and as it happened, I didn't have my wire cutters with me. I had to borrow the farmer's and they weren't up to the job, so I had to sever the wire strand by strand from around her neck. Twice the sow slipped out of the catcher I was using to restrain it and twice she bit me. Fortunately, sows tend to snap and let go, but boars, as I also know to my cost, will bite and hold on. Even so, the cut was deep, I almost lost a finger and as soon as the sow was freed, I was on my way to hospital. It was the first time I had released a badger this way. Normally I manoeuvre them into a sack or wooden box and then, and only when they are secure, do I cut the anchor wire and take them to a vet to remove the snare. The difference in this instance was that the sow was lactating and her udder was tight and swollen with milk, so I was eager to release her, for the cubs would not have suckled for at least 16 to 18 hours. I took a chance and she bit me. It was my own fault, but at least I had the satisfaction of seeing her race away in the direction of the sett and back to her cubs.

At this point, I would like to repeat the warning you'll find elsewhere in this book: don't try to handle injured or cornered badgers; they are powerful animals and their jaws are those of a scavenger – strong enough to crush bones and to tear through tree roots. The instinct of a frightened, injured animal is to bite, so don't put yourself at risk. Get help – call the RSPCA, the police or a veterinary surgeon.

Road victims: what you should do

Sadly, most people see badgers only as carcasses on the roads. Thousands die that way every year, but some badgers are injured rather than killed and many can be saved or helped. So, what should you do if you find one injured? Note very carefully where you find it, use a tree, a local landmark, an oddly shaped bush, anything that will help you locate the exact spot again if you do have to drive off to find help. That could save hours of searching, as I have found to my cost. It will also help when, hopefully, the animal has recovered after veterinary or rehabilitation care and needs to be returned to the wild.

Most badgers are injured at night, of course, which is when they are often found. So, that's the first problem; vets are unlikely to come out in the early hours of the morning. The nearest RSPCA inspector may be 40 miles away and not available. If you are lucky, you may know someone in the local badger group or in a rescue centre who will respond to an early-morning call, but you may have to take care of it yourself until the next day.

So, here are a few tips that should help. Let the animal know you are there, but park at least 50 m away, walk slowly and talk quietly as you approach. Remember always that an injured badger is a frightened badger, a dangerous badger. Do not handle! If it appears to be unconscious, try the stick test: touch it lightly with a stick, never with your hand, to see if it responds. If it is conscious, do not try to push it head first into your box or carrying cage, but try to make it go in backwards. A cornered or anxious badger, almost any animal in fact, will attempt to retreat. It won't willingly go forward. If it is well enough, it is much more likely to resist, to dig its feet in and to use them as an anchor. So, stand in front of the animal and position the box behind it. If you have someone to help then so much the better. If you're alone, try to use the contours of a bank or a hedge to encourage it to back away from you into whatever container you have. A sheet of newspaper, or your coat can be a help. In 50 years of pig farming, I have dealt with many dangerous boars (and believe me they can be killers) by simply holding a sheet of newspaper or a paper sack in front of them. They back away, often directly into the carrying box or transporter you want them in, because to them the paper is an impenetrable wall or obstacle; they don't realise they can go through it and assume they have to go round it. This prompts me to ask why huntsmen (when fox hunting was legal) didn't use newspaper, rather than clods of soil, to block sett entrances on the day of a hunt. It would stop the fox getting into the sett, for there's no way he would try to bolt through it. Furthermore, the paper would soon disintegrate with rain or, if it remained dry, the badgers would be quick to pull it in as bedding that evening. If animals can't see, they won't go, but put some wire netting up and they will try to go through it.

Back to our injured badgers on the roadside. Once the animal is safely in a carrying box, leave it in the box, preferably with plenty of hay or straw bedding or old clean clothing, until you can get it to a vet. That may not be for several hours, but at least it will be warm, dry and quiet. If you have to take it home, don't take it out of the box. Simply leave it where it is in the box, in your car or, better still, in your garage where it is dark and quiet. That way you minimise stress to the animal. Remember, at all times that it will be frightened and that it has a fearsome bite. Do everything you can to keep stress to the minimum.

Road casualties are one problem you may encounter: snares are another. Cruel, indiscriminate and totally unacceptable to anyone who cares about animal welfare, snares are most often set to trap foxes and rabbits, but frequently they catch badgers, cats, dogs, lambs, sheep, deer or cattle instead. I have had to rescue far too many. Some do manage to escape or break free, and many of the road casualty badgers I have examined have had snare marks on their bodies – the flesh had healed, but parts of the body had been permanently disfigured. One had deformed ribs, and it was obvious it had been caught as a youngster and the snare had rusted away months later. Snares have somehow become accepted, but think of the reaction if a horse was tethered by a wire and left for several days in the blistering sun with no food or water. The public would be outraged and the perpetrator would be sent to jail for years.

Whoever is faced with the job of freeing a snared badger, there is one golden rule, in my view, that must be followed. First, get the snared badger into a box or sack. That will help settle it down and it will be secure. Then and only then, should the snare wire anchor be cut. Read the earlier section about the lactating sow that I set free, if you want more convincing. I did everything that day that you shouldn't do, and I paid for it with a badly severed finger. I must explain that, had it been a boar, or a sow without dependent cubs, I would have put it into a box and taken it to a vet to remove the snare under anaesthetic, but I felt I had to get her back to feed her cubs, so I took a risk and paid the price.

One last point about snares. I have freed many badgers from them. Some (the legal kind) are supposed to be free-running, but I've yet to find one that works that way with badgers. Snare wires become so twisted and distorted as the badger tries vainly to escape that they won't slacken off enough even to get the wire cutters underneath them. There is, in my experience, no such thing as a free-running snare, once an animal has been caught in one. The snare has to be cut free, and most often that can be done humanely only when the badger has been anaesthetised. But why, why, why are snares still legal?

As an aside, I have to say to anyone thinking about rehabilitating animals that you have to be prepared to be on call night or day. I remember one morning when we were woken at 2.30 am by the telephone. It was West Mercia Police. The message: "A constable is on his way to you with an injured badger." I dressed, put the kettle on and waited. The constable duly arrived with the badger that he had found on the roadside entangled in a snare that he had removed, but the badger had one nostril severed and obviously needed veterinary attention. I made it as comfortable as possible and we went in for a cup of tea. During our conversation, I asked why he had not taken the badger to the vet. "I didn't like to wake him at this time of the morning," he told me. No, I thought, but it was OK to wake me!

A badger in a sack and the owl that wheezed

Anyone involved with the care and rescue of wildlife will know only too well that you have to try not to get too emotionally involved. There are plenty of moments when it would be all too easy to become depressed – the sight of a strong, fit animal in a snare, the savaged remains of badgers victimised by baiters, the mutilated carcasses of foxes, the crushed bodies of road accident victims, all are disturbing to say the least. But, you have to try to be as calm and detached as you can, and even the grimmest incident can have its moments of light relief.

Some years ago, the police rang me to ask if I would help release a badger in a snare. I agreed to go, but insisted that a police officer went with me, as I had no right to be on the land in question. In fact two police officers came along. When we reached the farm, we had to walk half a mile up a very steep hillside. I didn't take my carrying box, but I did have a sack and that suited me, for experience has taught me that an animal in a sack has a certain amount of comfort. The RSPCA don't like it, but an animal carried that way never injures itself, never fights and generally settles well. It's the darkness around it that does it and the comfort of having something soft and warm against the body that helps alleviate the stress of the occasion. In this instance, we found that the badger had been caught around the neck by a snare placed between a wall and a barbed wire fence. In its efforts to get away, it had become entangled in the barbed wire; we found it hanging with its front paws in the air. And, thank goodness they were, for underneath the badger, a few inches from its paws, was a large stone, close enough for the animal to reach with its back legs as it struggled. The claws had been worn right down to the quick and had it not been for the way it was hanging, the same would have happened to the front claws. A badger's front claws, of course, are absolutely vital, not only for helping it to feed, but also for excavation, grooming and for defence. Happily, all its other injuries were superficial; the snare had cut the skin, but the wounds weren't very deep. I eased the badger into the sack, tied the sack firmly with the snare still around the badger and then, and only then, was it safe to cut the snare anchor chain. As mentioned earlier, never cut the anchor until the animal is secure in a container, as it could escape, still entangled in the snare. Very casually, I slung the sack over my back and headed down the hill, the police grinning as we went and one observing that, if they saw me beating the 4-minute mile, he'd guess that I'd had a bite on the bum from a badger. They thought it was funny. I wasn't so sure; I know what a badger bite feels like.

We headed for a veterinary surgery. There, the vet anaesthetised the badger, removed the snare, examined the wounds and said that, although they were nasty, they ought to heal in a day or so. We talked over what to do and agreed that, as soon as the animal came round and got its wits together, I would release it. That meant letting it go in the middle of the night, but no matter. We took it back to exactly the same spot, this time in the company of a presenter from Radio Shropshire who wanted to record the incident live. As we struggled up the hill, he said, "Listen to that, George, I can hear owls calling." I stopped, listened, but heard nothing. So, we started on up the hill once more. "There, there it is again," whispered my earnest companion. I stopped, listened and grinned. Owls, be blowed, his recording equipment was picking up the sounds of George Pearce wheezing! Carrying a full-grown badger plus a box (combined weight 28 kg) up a Welsh hillside takes some doing for an 'old un', especially one who suffers with asthma.

Once we were back at the spot where the badger had been trapped, I opened the box and stepped back. The badger was in no hurry. It casually walked out, lifted its head high and sniffed the air, just as badgers do when they emerge for the first time at night from their setts. Then, it slowly ambled off back home along a well-worn badger path, stopping only once to glance back. All the badgers and foxes I have released, once they know they are back on home territory and have put some distance between us, have done exactly the same thing. They stop, turn around and look towards you as though to say, "Ta". This one was no different. It knew where it was and it knew where it wanted to go. Walking back to the car, it was my radio friend's turn to puff a bit – my load was lighter than his now!

That was a successful badger release: a freed animal returned within a few hours to its own patch, its own social group. In a sense that badger was lucky. As far as we could tell from its injuries, it hadn't spent too long trapped in the snare and its wounds were such that we were able to release it quickly. Others aren't so fortunate. I have very fond memories of that badger and I'll never forget one young badger snared in a neighbouring county. It was mid-summer and the country was enjoying a heat wave. I was carrying out a survey and found a young female badger in a snare. The vet estimated that she had spent at least four agonising days in the snare before she was found. In all that time, she had no water to drink, no food, no shelter from the blazing sun and her flesh wounds were crawling with maggots. This was cruelty at its worst. We got her to a vet just in time, for she was only hours away from death. Her wounds were stitched, she had to be weaned back on to food and when she was back on her feet, they gave me a call. Would I take over? Of course, I said yes.

When the badger arrived, I placed her in a rehabilitation pen, opened up the carrying box, stepped back to allow her to come out, and then looked on horrified. The badger walked like a crab! Soon I could see why: to repair some of the damage caused by the snare, the vet had taken some skin for grafting and although the wounds had healed, the skin was tight in places, too tight. The badger was a sorry sight and I must admit that I questioned whether the vet had been right to try to save her. But nature is a wonderful healer and some animals have a tremendous will to live, as this one had. She was a real character. Just four days after she had hobbled out of her box with her odd crab-like walk, where did I find her? Clinging to the underside of the wire roof of her cage doing her damndest to get out. The vet was right: he had seen how determined this little sow was to live. Her determination and willpower was an inspiration, a wonderful example of how the most serious injuries can be overcome. She deserved to live and in no time at all she was ready for release.

But where? Her old home territory wasn't suitable and there are few areas in Shropshire without badgers. But, I recalled that a young man who was keen on badgers had an uncle, a local retired farmer in his eighties and he had a couple of disused setts in his wood. So, he took me to see the old boy. We knocked on the door and explained what we wanted. His opening response was none too encouraging, "What on earth are you doing bringing me a ******* badger. Why would I want one of them damn things on my land? All they do is dig holes all over the place," he grumbled. I turned to go. No point in arguing with this old beggar I thought to myself. Then he stopped me. "Just supposing, just supposing, mind you, that I said yes, what would I have to do?" "Next to nothing," I assured him, "Just keep an eye on the sett from time to time. We'll provide a bit of dry dog food to leave nearby and if you could put it out some fresh water and a handful of straw now and then for bedding, there's nothing else she'll need." He glowered and grumbled, "Oh, all right then, I suppose," adding quickly, "But that's all, mind. I must be off my rocker."

We took the badger over to the wood in her carrying box, but before releasing her I took some of her dung collected from the pen and smeared it in and around the sett and pushed a bit of her old bedding into the entrance. At least she'd be greeted by a few familiar scents and it might feel a bit more like home. With that, we released the badger and walked back to the van, hoping for the best. A couple of days later, we got a call from the nephew. "The old boy's like a dog with two tails," he chuckled. "Can't keep him away from the sett, he keeps ringing me up to tell me what he's seen – it seems as if everything is going just great." I had to smile, not only had we found a guardian for that badger, but her strong character had also made an unlikely convert, for that old farmer in his younger days had been the master of a local fox hunt.

Another badger we were asked to look after was a road accident victim, a boar. He arrived with a steel pin that a vet had inserted to help repair a broken leg; he had lost a few front teeth and one of the claws of his left forefoot. An ear was also slightly damaged and crumpled, probably from an earlier fight, rather than from the road accident. Two or three weeks later, one Sunday morning, I had a call from the police asking if I would examine a badger sett, where they had arrested five men for suspected badger-digging. They arrived in a police car to collect me and as we neared the scene, it occurred to me that we were getting very close to the spot where that latest badger had come from. The person who had found him on the road was also the one who had spotted the badger diggers at a sett nearby. As soon as we arrived at the sett I began my assessment, looking for the clues that prove current badger use and as I worked I listened to the police. It soon became all too apparent that the farmer whose land we were on had been none too helpful; he had threatened the police with violence and had done his best to keep them off his land. Well, to cut a long story short, the case went to court and all five men were convicted.

Eventually, the badger was ready for release, the leg had healed and an X-ray had shown it was sound again, so there was no reason to keep him any longer. But, of course, it was now clear that there was no way the badger could be returned to that sett; it simply wouldn't be safe for him or for me. So, where could he go? Man had caused the injuries, so I felt it was up to us to help him recover and return to the wild. It's not something I like to do and I had not done this before, but I decided to let him loose in my own patch. By now, he had become used to the smells and sounds of our farm and our own wildlife garden, created over the years with loving care and skill by my wife Cris, and was content enough to roam about in there. The main local badger sett with its own established social group was 200 m away and I felt that, if he ran into trouble with those badgers, he would at least have the security of the pen, impregnated with his own scent. We left the pen open, but it was two or three days before he actually ventured out. I soon found that he had dug a tunnel at the back of some of the pig buildings and enlarged an old rat hole. I put food down each night and he stayed there for about a week. Then I realised he had moved on. 100 m away was an old rabbit warren and a few days later I noticed that one part had been opened up. He stopped there for about three weeks and then I lost track of him.

The following year, a neighbouring smallholder (a retired police chief superintendent) rang me with the news that he had a dead badger in the middle of one of his fields. As the crow flies, that was just under a mile away. I went over to look at it. The badger was covered in mud and all round it were hoof marks. The smallholder said to me, "I'm sure it's my donkey that's done him in. When he's in a mood, he rears up on his hind legs and comes down using his front hooves as weapons. I've had to rescue my dog from him before now." Well, I looked at it and pondered. My first reaction was that the badger was a road accident casualty that had crawled away to die in the field, but he was covered in mud and those hoof marks were everywhere; perhaps he had, indeed, been attacked as he foraged. As I looked and wondered, I saw the front left paw had a claw missing and the left ear was crumpled. It was the badger we had released more than a year ago. In that time, he had been in the wars again, for his right ear was missing, not particularly unusual in badgers. There was something else different about this badger; he had no fleas, which would be usual on a wild badger, as he had been treated whilst in captivity. I decided to take this badger for a post mortem. The vet rang me the following evening. He said, "George, it's just as though he had been in the ring with Mike Tyson. There wasn't a whole bone left in his body."

So the smallholder was right, the donkey had killed the badger. It was an unlikely end and a tragic one, but at least I had discovered that badgers can integrate into the areas of other social groups. It appears as though this one occupied a subsidiary sett some distance from the main sett. He was also an older badger and as such no great threat presumably to the top boar of the group. It was unlikely that he would have challenged for dominance and had probably been tolerated. A strong, young boar would not have been so lucky.

Help them, but don't hang on to them

Those, then, are some of my memories of my shared days as a voluntary rehabilitator. I say shared because so much of the practical day-to-day caring was undertaken by my wife Cris and my children Tristam and Pamela, while I got on with running the pig farm. For our family, there were days of pleasure, disappointment, delight and occasionally grief. I learned a lot from the wild animals that we looked after and I would like to offer a few words of advice and some important basic rules of thumb, for anyone thinking of doing the same, or indeed to those already involved.

There's a tendency to keep animals too long in care. If their injuries are slight, just a few cuts and bruises, or if they have simply been mildly concussed, 48 hours or less is often long enough in care; sometimes an overnight stay is all they need. It is essential that they do not lose their characteristic social group scent, which is vital to them for recognition and acceptance within the group. That scent comes from mixing and mingling, and it is a phenomenon that most farmers will recognise, for exactly the same is true of farm livestock. Pigs and cattle, for example, if separated for a few weeks, will fight when put together again. Why? Because they've lost that previously instantly identifiable group scent. Captivity for wild animals is unnatural and it can be traumatic and stressful. As a general rule, the sooner they are released, the better it is for them. Keeping them too long is not a kindness – quite the reverse: often it is an indulgence, a case of the carer getting more out of it than the cared for. Knowing when to let go is a fine art and we must always put the interests of the animals first, which means releasing them just as soon as they appear to be fit and well and able to care for themselves. We all know how, after a spell in hospital, our recovery speeds up when we get home. The same is true of animals.

Broken bones can take a long while to mend, but wounds are different, they can heal extraordinarily quickly, in hours, rather than in days or weeks. What about the conditions that wild animals and birds are kept in? I'm not too keen on large rehabilitation units, as often they are too much like hospitals. I would rather see a large number of small units, that are, or should be, closer to nature. Wild animals respond to the smells and sounds of the open countryside: they need fresh air, the chance to get out into the open, to get their backs wet and to feel the wind through their hair. In short, they need to have the chance to savour the natural elements, but they also need the option of somewhere warm and dry to sleep.

A recovering animal needs two environments. It must have somewhere to lie, where it can be warm. It must also have somewhere where it can be cool. Given those two sets of contrasting conditions, with clean, fresh water available to it at all times, it will find and choose the temperature it needs. A constant temperature is wrong – think how wretched we get in such conditions. An open, secure wire-mesh pen attached to a well-insulated kennel provides the two conditions that a wild animal needs. The kennel will keep it warm, provided it contains plenty of dry straw or hay (preferably hay). The bedding must be fresh, it must be dry and above all it must not be mouldy. Mould and animals do not mix. Mould is caused by changes in humidity. An animal will only urinate in its bedding when its body temperature is too low, so we should treat this as a sign that its accommodation isn't up to scratch. Let me repeat that golden rule: two temperatures within one enclosure. Too often, it is overlooked, or not even thought about. Most farmers, and certainly skilled stockmen, will always give their animals those options. The wire pen does the job.

A couple of other useful husbandry rules for animals in captivity, one of which I've already mentioned. Talk to them in a soft, gentle voice, and talk to them as you near their pen or cage so that they know you're approaching. Talk to them, too, as you care for them and feed them. I know they haven't a clue what you're saying, but a warm, sympathetic voice does help and it sparks a positive response.

Secondly, don't fuss them too much and don't keep looking in every half hour – that's wrong. If you've done your preparations correctly, they will have all they need. Fresh water and the right amount of fresh food are two other essentials. With foxes, raw pieces of rabbit are first rate. With badgers, some mashed-up earthworms, soil and all, or a raw egg, placed in a corner where they can find it, are all they need at first. Worms are, after all, the most logical food to offer a badger, but you'd be amazed how few people ever think to try it. Cubs of one month to five weeks old don't need to be bottle-fed and as long as they have some moist food, they'll do well. Raw eggs are marvellous for all injured and convalescing animals, young and old. Provide the food, then leave the badger alone, don't disturb it, let it sleep. Look in each morning to check how it's doing. If it's asleep, it will almost certainly be covered in straw. Lift up the lid of the pen and listen to the sound of breathing. If the breathing is slow and steady, leave it alone until the next feeding time. If, when you return, the badger is reluctant to move, try to look at the eyes and the lips: the eyes should be bright, the lips pink. If you wait long enough or even offer a small stick to its mouth, you should get a glimpse of the lips: if they are very pale or dark red or inflamed, I suggest you have a word with a vet.

One other general rule is worth passing on. Adult animals – badgers, foxes, stoats, weasels, etc. – don't need the company of others as they recover from injury. Cubs do, they thrive in the company of others. If you have other youngsters of the same species, that's ideal, so let them mix and play. Young carnivores, in particular, need to exercise body and mind and are always very playful; that's how they train and prepare for life ahead. When our children were small, we used to let them play with the young badgers and fox cubs and we would also leave the cubs in the company of the cats and dogs. Later in the evening, I would get them out of their box and groom them. Fox cubs would enjoy every moment they were being groomed, but when they had had enough they would let you know. They would spring on to the back of the chair, on to the sideboard, television or window ledge, or make a grab for the curtains and cling on. Invariably, they would also deposit a small dropping on top of the table, the television or some other conspicuous place, a habit that mimics exactly what happens in the wild. No televisions or tables there, of course, so they choose stones, tree stumps or other equally conspicuous features, including fungi. I soon tumbled to what was happening to our captive cubs and I would put newspaper on their favourite marking places, and they would always use it.

It is very important, as I quickly discovered, for young carnivores to have plenty of play stimulation, from whatever source. As they mature, their behaviour and their needs change: they start to become independent, and ignore the things that previously interested them. For example, as the weeks pass, they no longer crave the company of other animals or children or grown-ups, and will tolerate only the person who regularly feeds them. At this stage, they are ready to go their own way, to return to the wild, and they need no training, knowing instinctively what to do.

Chapter Six

Badger consultancy

Ducks' legs and raindrops

Before going into details of consultancy work, and the responsibility that entails, may I quickly tell you about one of the most testing roles I have ever undertaken – that of expert witness in court cases involving alleged badger persecution. My introduction to the role began many years ago through contact with the RSPCA and the local police. Five or six of us within the badger group were picked out as possible expert witnesses and we received some very useful training; how to approach a scene of crime, for example, what clues to look for, what action to take and how to give evidence. Our training involved briefings by solicitors and mock courtroom trials.

It wasn't long before my training was put to the test. I was called to a badger dig in North Shropshire; five men had been arrested and the police wanted me to collect evidence. The men had smashed concrete to get under a shed where badgers were living. The case took over a year to come to court and the trial lasted 13 days. I had never been in a courtroom before and I was about to be given some of the best advice I have ever received. A senior policeman took me to one side: "George", he said, "as you go in, look around the courtroom, take a look at the legal counsel and remember they know more about the law than you will ever know.

Then shut that side of it out of your mind and concentrate on what happened at the sett that day." He was absolutely right; it was the best advice I ever had. I was able to detach myself from all the paraphernalia of the courtroom and concentrate on what happened. I asked how long I would be there in court. About three hours, he said. Well, I was very nervous. The prosecution questioned me for about three hours and then the defence started to cross-examine me. In all, I was in the witness box for seven hours.

I have described elsewhere (Chapter 2) how I was cross-examined about a badger path that led to a fence and how I was able to determine that it led under, rather than alongside, the fence. What is worth adding here is that this trial taught me what to expect from defence witnesses. One, a so-called expert claimed he had visited the sett recently and had found poultry feathers and the legs and feet of ducks there. Strange, that, I thought. He then went on to argue that I was mistaken and that the sett, where the alleged offence occurred, was in fact a fox's earth. To throw some light on the dispute, the magistrate ordered us to go back to the sett. Well, we went in our lunch hour and the defence witness pointed to what he said were the claw and paw marks of a fox. I identified these marks as pit marks caused where the rain had dripped from a tree and a police sergeant involved in the case, a very knowledgeable wildlife man as it happened, confirmed this. As for the chicken feathers and ducks' legs, which had been chopped off with a cleaver or similar cutting tool, they had been thrown there, to mislead. We won our case and two of the men went to jail.

Make way, make way

Like it or not, we live in a world of constant change. We cope because we have to, or because change brings with it clear benefits. For wildlife, however, the picture is much bleaker. New farming practices (many of them the direct result of the EEC's Common

Agricultural Policy) have, for the most part, been to the detriment of our birds and mammals – creatures that, for too long, we have taken for granted. They have suffered dramatically as the drive for greater efficiency on the land has gathered pace. Hedgerows have been ripped out, trees felled and ponds and wetlands drained, all through EEC subsidies. These changes would not have happened to such a large degree had it not been for these subsidies. Old pastures, too, have been ploughed up, pesticides and weedkillers used in ever-greater quantities and planting seasons have changed, factors which have all had a detrimental effect on wildlife.

But, of course, it is not simply the farmers and the landowners who affect the well-being of our wildlife: we all do. Most of us accept the need for new houses, more factories, better, bigger, wider roads and new or improved gas, electricity and water supplies. All of these either take up more land or impinge on what is already there. Wild animals (badgers in particular) often have to move out. Put simply, they are deemed to be in the way. And, as if that were not enough, the badger has long been singled out for persecution of the worst kind: digging and baiting, and many have also been shot, snared, gassed or poisoned.

Those, then, are just a few of the reasons why the badger needed more protection under the law.

Protection of the animal came first. The badger has been a protected species since the Badger Act 1973, and has enjoyed additional protection under the Wildlife and Countryside Act 1981, and a later amendment in 1985. But, simply protecting the animal was not enough, as continuing acts of persecution and destruction demonstrated all too clearly. A mammal that spends more of its life below ground than it does above ground needs protection as it sleeps, rests, breeds and brings up its young. That was a battle that took a long time to win, but it was achieved in 1992, thanks mainly to the work of the League Against Cruel Sports, Wildlife Trusts, the National Federation of Badger Groups, the RSPCA and the Vincent Wildlife Trust. They worked together as the Badger Coalition and

1. A typical spoil heap. Note the steep sides due to badgers dragging soil out of the sett.

3. Badger sett entrance. Note the polished sides and roof caused by regular use.

2. (Top right) Typical badger sett entrance. Note the fresh excavated soil and badger paw prints.

4. (Right) Green bedding collected by a sow to keep her cubs' chamber warm.

5. A hollow ash tree used as a sett by badgers on a flood plain.

6. A badger cautiously checking for danger before leaving the sett.

8. Badger emerging in snow.

7. (Top left) Two badger cubs at a sett entrance. One is emerging cautiously and the other is grooming.

9. (Middle left) A badger startled when I stepped on a twig.

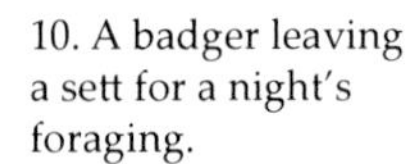

10. A badger leaving a sett for a night's foraging.

11. Badger in a common foraging position. Note the long neck which is typical of the weasel family.

12. Foraging in gardens is more common than people realise.

13. A badger foraging in the rain.

14. Two badgers fighting.

15. A badger with both old and new fighting injuries.

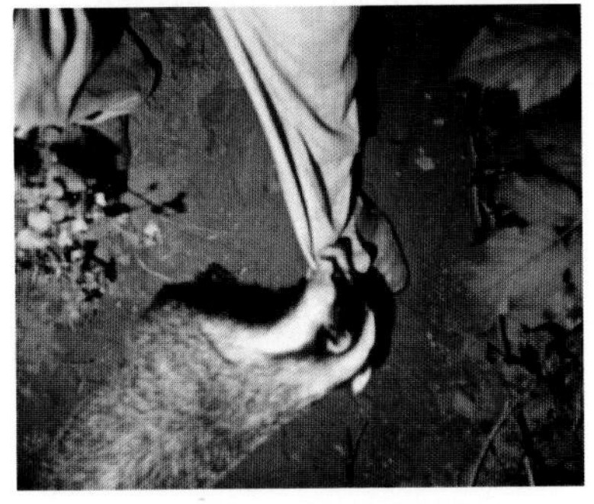

16. A badger attempting to use the author's trousers as bedding!

17. A badger reversing out of a sett, bringing out soil to extend the sett.

18. A badger using a latrine.

19. A badger paw print.

20. A badger scent marking the author's foot!

21. A badger exclusion gate.

22. A badger latrine used by two different badgers. One had eaten earthworms - the resulting faeces is soft and runny. The other had eaten peanuts - these are clearly evident in the faeces.

23. An exposed badger chamber with bedding.

24. Inside a tunnel system, 3 m down into the sett.

25. The end of a tunnel during excavation deep down into the sett.

26. Badger hair caught on barbed wire.

27. The more common black & white badger with the less common erythristic.

28. An old badger track, still in use after the field has been ploughed.

29. Badger paw prints in the snow.

30. One of the thousands of badgers sadly caught in snares every year.

31. Fox and badger cubs with a Vietnamese pot-bellied pig during rehabilitation.

32. An erythristic boar.

33. Nine of the eleven 'flood badgers' asleep in a ball.

34. One of the 'flood badgers' asleep in a tree.

35. A young cub getting a taste for peanuts.

36. A badger asleep in a rehabilitation chamber.

37. My daughter Pamela badger-watching as a child.

38. Hiya the enchanting fox cub beginning her rehabilitation.

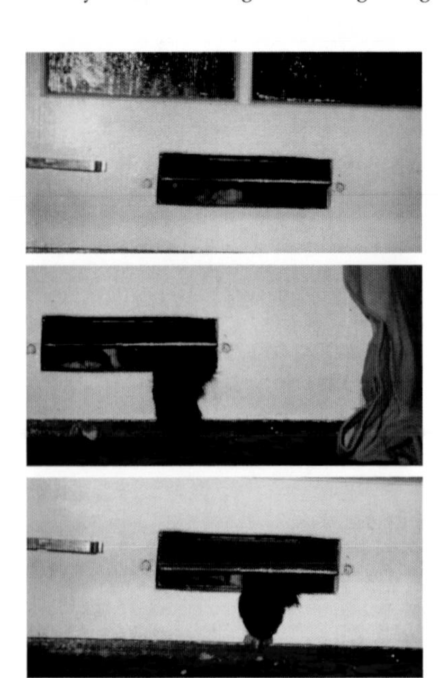

39. I smell food.
40. I know it's here somewhere.
41. Got it!

42. The author with Bodger on his arrival.

persuaded the Government to bring in the Badger Sett Protection Bill 1991 (subsequently amended in 1992 to The Protection of Badgers Act), which consolidated previous badger protection measures.

Today, as a result of their campaigning, the law states that a person is guilty of an offence if they: (a) damage a badger sett or any part of it; (b) destroy a sett; (c) obstruct access to or any entrance of a sett; (d) put a dog into a sett or; (e) disturb a badger whilst in its sett. Licences to carry out work on, or near, a sett can be applied for from the appropriate statutory body. Unfortunately, there are many business people, farmers and landowners, who don't like either the laws that protect the badger or those safeguarding the sett. The result of this hard-won legislation today is that the badger's welfare has to be taken into account whenever any kind of development is planned that would threaten either the badger or a sett. In practice, this means that, if an active badger sett is close to new building developments, new roads, new railways, gas, water mains or sewage pipes, even new extensions to houses and buildings, that work cannot go ahead until a licence to 'interfere' with a sett has been obtained from the appropriate licensing authority.

There are some exceptions to this protection: DEFRA officials may trap badgers and close setts, and farmers are allowed, as part of their necessary work, to plough, sow and harvest land containing setts. But, in general, setts now have a high degree of protection and penalties for breaching the law include substantial fines or jail sentences. Not only is it against the law to wilfully kill or injure badgers, but it is also an offence to disturb them in their sett. Difficult both to define and prove, that final safeguard is none the less an important one that to be taken very carefully into account when it comes to making decisions about what to do with badgers when they are 'in the way'. That's part of my job and one I take very seriously, as I explain in the rest of this chapter.

There are claims that, since the new sett protection laws came into force in 1991, badger numbers have increased significantly

– 'explosion' is the term often used to describe the change. I disagree. What we have seen, in my opinion, is a marked increase in the number of badgers on livestock farms and a reduction on those that are chiefly arable. The explanation is not hard to find: farming practices have changed for the worse as far as the badger is concerned and again the rules, regulations and subsidies available through the EU are chiefly at fault. For some time, the trend in the UK has been away from livestock in favour of arable and pastureland has been ploughed up everywhere to make way for crops. Arable land is a less favourable habitat for the badger. Put simply, it provides much less year-round food, so the badger has gradually moved in greater numbers to the remaining areas of grassland. The increase in badger numbers is, in my opinion, more perceived than real. It should be remembered that there has never been a badger count, only a sett count.

The law, and the protection afforded to the badger and its sett, need to be applied with commonsense. That way, long term, the badger is the winner. Impose the law in a rigid, bureaucratic way, impose needless rules and costs, and people will begin to take the law into their own hands again. If that happens, the badger will be the loser. In the pages that follow, I spell out in a more practical way some examples of good and bad conservation as it currently affects the badger.

Have they read the book?

Developers rarely have the kind of expertise needed either to recognise or deal with the presence of badgers on sites earmarked for development, which is why they rely on others. More often than not, full-time ecologists or badger consultants are called in. To do the job well, whoever is chosen should have the experience to decide what is needed for the long-term survival of the badger and just as important, have the time to see the solution through to its conclusion. Sometimes, it is a relatively simple job to protect the badger, while allowing the proposed development to go ahead, but often it is not. It may require weeks of observation, planning and sheer hard work to create a solution that is both fair to the badger and acceptable to the developer. Occasionally, the only way is to close a sett and move the badgers elsewhere, possibly into a man-made, artificial sett, but my preferred and prime objective is always, if at all practical, to leave the badger and its sett where it is.

Badgers, as I demonstrate a little later, are surprisingly tolerant. They will put up with all sorts of noise and disruption, quite happily learning to live with fairly radical changes to their immediate environment, provided that they continue to have access all the year round to adequate food supplies, are able to live their lives free from persecution by people and free from harassment by dogs. In my role as a consultant, the welfare of the badger always comes first, but the needs of the developer have to be recognised and met wherever possible. So, I look for practical, commonsense solutions that work for the badger and don't involve the developer in unnecessary expense. Read on, and I'll tell you about badgers that now live contentedly, cheek by jowl with all sorts of new buildings, roads and railways, even airfields. The sights and sounds around them have changed, but two things have remained in their favour: safety in their setts and the availability of plenty of food within reasonable distance.

From here on, as convenient shorthand, I shall use the word 'developers' to describe those who plan work that impinges on badger territory. Often, they are developers as we normally understand the term – builders mostly. But large public companies, the Highways Authority and public utilities figure prominently in the casebooks of badger/wildlife consultants, who are invariably self-employed, and full-time ecologists who are either self-employed, perhaps as part of a consultancy with a wider remit, or on the staff of a council or public body. Ecologists, of course, deal with a whole range of wildlife and conservation issues. Some are very knowledgeable about badgers, others less so. That is why my clients include ecological consultants; they commission me for aspects of badger work that they feel insufficiently qualified to take on with confidence. With the exception of emergencies, developers have to plan their work so that it takes account of protected wildlife, Sites of Special Scientific Interest (SSSI) and so on, *before* the work starts, while it is being carried out and after it has been completed. The *before* bit is very important.

We have all heard of the earth-moving equipment that goes in and flattens everything in a short space of time, or the trees that are felled just as quickly, or the listed buildings that are destroyed before anyone has had chance to intervene. And, I guess we are all equally unimpressed by the predictable excuses and expressions of injured innocence that invariably follow. That still happens, unfortunately, but the first task for responsible developers is to carry out a survey to establish what plant and animal life is at risk, or is potentially affected in some way, by the planned changes. They have to take account of rare and protected species – anything from orchids, to great crested newts, the nesting sites of peregrine falcons or the setts of badgers. Quite right, too, for all wildlife is important and if we are destroying their habitat, we are also destroying our own.

Badgers, as we know, do not ask for very much from us; they simply want a place to live, somewhere underground that is safe, dry and warm, and somewhere to feed. Number one on their menu is the humble earthworm, so that means their preferred places to live are those where worms and other tasty morsels like beetles and grubs flourish. That puts established pastures at the top of their list, with hedgerows, woods, copses and grassy and shrubby areas, public open spaces, golf courses, parks and playing fields of almost any kind (including gardens) not far behind. All of these areas are subject to the pressures and the intrusions of modern-day life. Today's open field or derelict factory site is tomorrow's building site. That undulating piece of countryside with its hedges, woods and gentle sloping pastures may be tomorrow's new ring-road, bypass or dual-carriageway. That fast-expanding town or village is the next to need more public services – water, sewerage, gas or electricity. Time was when those developments would go ahead with little or no regard for the badger and its habitat and I shudder to think how many setts have been destroyed, how many badgers, suckling cubs and mature sows and boars, have died under tonnes of earth as the bulldozers have moved in. That ought not to happen now, thanks to the Protection of Badgers Act 1992. If it does, and provided it can be clearly proved that setts have been unlawfully destroyed or badgers injured or killed, then developers face penalties of up to £5000, plus six months' imprisonment for *each* sett affected.

Statutory bodies issue guidelines for developers. A Natural England booklet called *Badgers and Development* spells out developers' obligations. It reads:

The Protection of Badgers Act 1992 is based primarily on the need to protect badgers from baiting and deliberate harm or injury. It also contains restrictions that apply more widely and it is important for developers to know how this may affect their work. All the following are criminal offences:

- *to wilfully kill, injure, take, possess or cruelly ill-treat a badger;*
- *to attempt to do so; or*
- *to intentionally or recklessly interfere with a sett.*

The Protection of Badgers Act 1992 explains what a sett is thus: 'Badger sett' means any structure or place which displays signs indicating current badger use.

Sett interference includes damaging or destroying a sett, obstructing access to a sett and disturbing a badger whilst it is occupying a sett.

Sounds straightforward enough, doesn't it, but that little word 'current', just like the word 'active' when referring to a sett, has been the cause of many hours of discussion, debate and argument for badger consultants and for the legal profession.

The definition of sett is interesting. You would be surprised how many people (farmers included) fail to recognise one. And, it is also worth noting that even reasonably experienced wildlife watchers, as well as some who claim specialist badger knowledge, fail to distinguish fox earths and over-sized rabbit burrows from setts. Part of my job, as I indicated earlier, is to be able to do just that with absolute confidence. A badger's sett is protected, even when a fox is in occupation; a fox's earth is not protected. Eight times out of ten, it is relatively easy to identify a sett as I explain in Chapter 3. Life for the consultant is relatively easy when the badger has read the reference books and conformed to all the classic behavioural patterns! It is rather less simple when they don't conform, or when the ground conditions are such that the clues are much less clear. Little wonder, then, that developers call in consultants to carry out surveys and give advice.

The first stage in any survey is to determine whether or not badgers are present. Large, active setts with fresh spoil, lots of badger paw prints, nearby latrines and badger hair in the spoil are the easy ones to identify. The smaller, scruffier setts with fallen leaves and hard-packed soil in the entrance and few other indicators of current use, or with a mixture of conflicting visual clues, are more difficult. Often it is necessary to monitor these day and night over an extended period. Straightforward night-time surveillance will sometimes provide the answer, but as I have already indicated it is possible to go several nights without seeing one badger emerge, so night surveillance often has to be supported by entrance monitoring. Badgers aren't exactly ballerinas; they lumber along in their purposeful, short-sighted way on large paws, their low-slung bodies brushing aside minor impediments as they enter or leave a sett. A word of caution here, though: signs of activity at or in the tunnel entrance do not prove that animals have *entered* the tunnel system and used the sett.

A few small sticks or twigs, pencil length, pushed firmly into place, close enough together so that badgers moving in or out of the setts will displace them, act as very effective exit/entry indicators and cause the badgers no concern whatsoever. When I am doing survey work, I put these in place during the day and check them the following morning. If I have reason to suspect that there may be the possibility of human interference to create false clues, then other methods, which I will not detail here, are used. A late-evening scattering of fine soil, or soil and sand just inside the sett entrance also helps to determine whether badgers, foxes or rabbits have moved the indicators. A badger is relatively heavy, of course, so in light sand or soil it leaves firm prints, quite different from those of a fox or a rabbit.

Years of observation teach you all sorts of things about badgers and also a lot about human nature. For example, from time to time, badger hair is placed at the entrances to disused setts by people who would like you to believe that they are, in fact, active. Thankfully, they usually lack the wit or the knowledge to do it expertly, and it is only too obvious that the hair has not been shed naturally. So much then for false clues.

Some of the things I am asked to do

To help you get an understanding of how badger consultancy works (and remember it's all about badger protection), let me sketch out just one broad-brush example. Then I shall deal with some of the underlying principles, before detailing some case histories that demonstrate the things that can be done to accommodate the badger in our ever-changing landscape.

The first thing that happens: the phone rings, it is, let's say, a local council's Highways Department. "We have a problem with badgers. Can you help?" The caller goes on to explain, "We've subsidence in the road. For years we've been patching it up and now we realise there are a series of badger holes going underneath."

My first visit is normally free of charge. I establish if the problem is indeed badger related and then, if it is, what can be done. The first thing I ask myself is: can the badgers stay? Is it feasible? I always ask for the engineer to be at the first site meeting as, until I have talked to him, I don't know exactly what structures are involved or what underground cabling or pipes may be at risk if the badger tunnelling is extensive. With roads, bridges, pipes and so on I need to know what they can tolerate in terms of change or interference. If the badger diggings are dangerous, a risk to public safety, then a decision has to be made quickly. It doesn't follow automatically that a sett undermining a road has to be closed. It is possible for the badgers to stay where they are once some innovative strengthening has been carried out.

Proposals to build houses or large structures require thorough survey work and constant monitoring. If, when I'm called in, I find one or more setts on land earmarked for development, then I like to survey all the area within a minimum of three-quarters of a mile radius of the site. That way, I have a feel for how much foraging is available, how good it is and to what degree the badgers will be affected, both during the work and when the planned development has been completed. If badgers are present on land designated for building, it is almost inevitable that the developers will prefer to have them moved. But I always go with an open mind and always look for ways of leaving the badgers where they are, if that is practical. Understandably, developers are loath to hold up work, and reluctant to accommodate the needs of the badgers by changing the proposed layout, perhaps with the loss of some houses. There are always solutions that suit both the developer and the badgers. Sometimes, the developer has to make concessions; sometimes the badgers have to be moved. On one site, for example, I created a small tunnel, which allowed the badgers to come and go as they pleased and the developers were able to proceed as planned.

There are some who think that badger consultants, because they make money from giving advice and carrying out surveys, are there to do what the developer wants. That most definitely is *not* the case and that's why I repeatedly refer in this book to the need for the consultant to be absolutely independent. They are there to protect the badger; they do not initiate development, that's the decision of others; they respond to it. I play it as I see it, always putting the safety, security and long-term survival of the badger first.

Main setts are the focal point around which so much of the life and survival of a social group revolves; the place where, for most of the time, the dominant boars and sows live, rest, sleep and breed. Annexes, subsidiaries and outliers aren't usually so important that the loss of one or two has a dramatic impact on the badgers. Observe how quickly and how often badgers dig new setts and you'll see what I mean. So, if the badgers can stay in the main sett, perhaps with the added protection of fences, bunds or walls, then that is the solution I go for. For that to be the preferred option, it follows that the badgers must retain, or be provided with, protection in the new development long after it has been completed and the developers have walked away, their job, their role, finished.

There are some exceptions to that last rule. Retaining a main sett, perhaps on the edge of a large new development, may in itself be a compromise, an interim measure. It may be apparent that, at some stage, the development will grow and expand, so that the sett becomes untenable or has to be closed. In such circumstances, it doesn't follow necessarily that the badgers will have to be moved out forcibly, they may choose to go of their own free will, they may look for, and find, a better, quieter site. The sett that's causing the potential difficulty may be abandoned completely in favour of another sett (that happens from time to time to setts unaffected by development, way out in the countryside), or they may be used only occasionally. During the past year alone I have recorded three instances, in different parts of the country, where setts have changed status. The main setts have been deserted and subsidiary setts have been taken over as main breeding setts.

Protection is a two-sided coin, of course – the badger often needs it, but so do people. Badgers in the garden would be a dream to some and a nightmare to others, a very real threat to their pristine lawn, their prized vegetables and their carefully nurtured plants. That is reasonable and understandable. We are not all wildlife devotees, but what must be remembered is that badgers roamed this land for thousands of years before gardens were even thought of. Just as badger setts that are close to, or are part of, a development need safeguarding so that they are largely left undisturbed, so too the owners of new properties adjacent to setts need the reassurance that badgers can't simply wander in and raid their gardens. Fences of the right height and construction are usually the best and least expensive option. Walls do a great job, the depth of their foundations discouraging badgers from tunnelling underneath, and it is easy to build in access and exit holes. Chain-link or close-boarded fences are more commonly chosen, as they are cheaper, presentable and effective. The wire has to be strong enough to resist the claws and teeth of a badger and the height sufficient to prevent a badger clambering over the top. Most importantly, badger-proof mesh has to be sunk well into the ground to prevent the master tunnellers from surfacing in someone's prize petunias. I was horrified to hear of a consultant and badger group planning to drive in iron rods at a distance of 10 cm apart instead of using chain-link fencing. They did not appear to know that a fully-grown adult badger can get through a 10 cm square. Fortunately, they abandoned the idea when I pointed it out to them. Mesh is invariably the best answer. Well sunk, it can keep badgers safe from unnecessary interference and people safe from unwanted badger intrusion, though not from the badger that decides to enter via the main driveway.

The design, layout and construction of 'badger protection

zones' (areas where the chief preoccupation is the safety of the badgers) vary from site to site. But, there are a number of basics I like to include. Provided there is sufficient room, I like to create at least a 30 m exclusion zone between the sett and the nearest boundary point of the development. This sometimes means that the developer has to allocate, to badger protection, a parcel of land that would otherwise contain one or more houses. Land is expensive, however, and the demand for new houses is constant, so sometimes setts have to be accommodated on much smaller areas of land, much closer to boundaries. It becomes a matter of judgement whether a sett, hemmed in on several sides, can survive as a viable home that badgers will accept. One sett had a protective boundary fence encroaching within 7 m, much to the horror of the local badger group, but I was able to satisfy myself and Natural England that the tunnels and chambers did not extend under the land being developed under licence. With my supervision, an exploratory trench was excavated and as no tunnels or chambers of the sett were found to be going onto the development site, a badger-proof fence was erected along this trench line. The badgers' preferred foraging was away from, rather than inside, the area being built on. At that sett, the badgers have raised cubs for three years in succession and appear untroubled by the proximity of the new fence and the 38 new houses. Indeed, they have actually extended the sett. In such circumstances, erection of the fence and all other building development work within 30 m has to be carried out under the supervision of a Natural England licence holder.

Protecting the badger on development sites is important, but we don't have to go overboard about it. I want to develop that point because, for the long-term future of the badger, it really is important that, in seeking to protect it, we need to strike a balance.

Like most animals, badgers don't need a lot of space to get from *A* to *B*. In the wild, badgers, foxes and deer routinely navigate their way through a jungle of natural obstacles, slipping through small gaps in fences, hedges, copses and bushes.

They repeatedly use the same narrow walkways and paths as they move from one feeding area to another and they'll often use man-made constructions like sills and bridges to their advantage. On one site I surveyed, the badgers actually walked along the top of a large diameter sewage pipe to cross a river, using it in effect as a bridge. In urban areas they do much the same thing. For example, badgers routinely visit gardens at night to forage, sometimes entering via drives, paths, alleyways or nondescript gaps between houses. Often, they'll simply slip under fences or ease their way through gaps in hedges.

Left to their own devices, wild animals are free-roaming, highly effective navigators. They go where they want, when they want, so why when new houses are being put up is there this pressure for wildlife access corridors large enough to drive a herd of cows through? Most often, I guess, the perceived need is in the mind of the beholder, the well-motivated but over-protective wildlife enthusiast who hasn't thought the problem through or who doesn't know enough about badger behaviour. It is most certainly not the case that bedding collection is a significant issue in an urban setting and often by choice urban badgers will use plastic shopping bags and crisp packets, even the odd old doormat or discarded string mop, as I have mentioned before. They have even been known to commandeer clothes off washing lines. Wildlife corridors, several metres in width, are entirely unnecessary if the aim is merely to allow access and exit. Different considerations apply if they are also intended to provide, or retain, foraging. I have already indicated that some housing developments can increase, rather than reduce, the amount of good foraging available for badgers and I would ask people to stop and think when they complain bitterly about a bricks and mortar project that they feel is depriving the badger and other wildlife of vital, irreplaceable feeding grounds. Some of the best wildlife habitats that I have encountered over the years have been in our towns and cities.

Habitat loss is not peculiar to towns and is not simply a by-product of development; it is happening all the time in the countryside too. To give you just one example: old pastureland is often put to the plough to make way for more profitable arable crops. When that happens, the immediate and long-term impact on wildlife is much more dramatic than if the same area had been turned over to housing development. Within hours, huge acreages of productive foraging disappear under the plough and if, as often happens, the first-year crop is potatoes, the badger loses out. Enormous quantities of artificial fertilizer, weedkiller and insecticides are applied and the crop provides nothing for it to eat. With housing, new turf is often brought in for lawns, topsoil is retained and a large variety of plants are planted in the individual gardens of the new home owners, so there is still foraging opportunity. Topsoil borrow pits, created while building is being carried out, also serve as well-stocked larders, not just for badgers but for other wildlife too. I am not in favour of allowing all developments, irrespective of their impact, and I am not saying it doesn't matter that land has been taken over for housing or offices, or the like. I am simply urging that we all keep it in perspective and that we don't insist on remedial measures that are either totally unrealistic or unnecessary.

Thankfully, we have largely moved on from the days when developers ignored the needs of badgers and other wildlife, either through ignorance or expediency. In my experience, with relatively few exceptions, they now accept the obligations put on them by the new laws and guidelines. Most conform to them and are likely to continue to do so, provided what's asked and expected of them is sensible, reasonable and justifiable. But, to insist on badly conceived, totally unnecessary and unduly expensive remedial measures isn't helpful and is likely to have an unfortunate downside. Some developers, exasperated by what they will see as unreasonable demands, may once again take the law into their own hands. The conservation cause will then suffer, and the badger will be worse off. So, we need balance, realism and well thought-out solutions to badger problems, and we need to ensure that the people who give advice have the practical experience, the commonsense and the knowledge that fits them for that role.

That's enough of the soapbox stuff, for the moment at least. Back to consultancy work and the role of local authority planning departments. They have to take a whole series of factors into account when deciding whether or not to allow a planning application to go through and I would not have their job for all the tea in China! The impact on wildlife is just one part of a complex equation they face in which all sorts of competing and contrary interests have to be considered and compromise is often the way through the maze as all sorts of objections are put forward. Their lot is often not a happy one. Some are really on the ball when it comes to wildlife, while others still have to be reminded of their obligations to check on the wildlife and conservation issues raised by planning applications. But, on balance, it is my experience that badgers nowadays get a good hearing and are viewed sympathetically, for the most part. Inevitably, planners are bombarded with complaints and, increasingly these days, people are likely to object on the grounds that the land to be built on contains badgers. Some do it for the best of reasons; others use it mischievously, simply as an excuse to try to prevent development that's in their own backyard.

The existence of a sett, or the fact that the land is prime foraging ground, is rarely deemed a sufficient reason for refusing an application, but it is now quite commonplace for planners, often at the instigation of Natural England, to require developers, as a condition of planning approval, to carry out work specifically targeted to preserve the long-term interests of the badger and other wildlife. Sometimes, the developers are required to accommodate the existing sett within the development area. Alternatively, they may be required to make provision to exclude them humanely from an existing sett or setts, and to provide alternative man-made setts close by. The knowledgeable, well-motivated badger consultant has an important role in all this. He has to look at the facts and put forward commonsense workable long-term recommendations and solutions.

You name it, they'll dig there

There's no telling where badgers will choose to make their setts. Provided there is ample year-long foraging nearby, they will set up home almost anywhere – under castles, pigsties, churches, hospitals, garages, sheds, barns, railways, main roads and canals, on military ranges, even on airfields, though happily these sorts of locations are the exception rather than the rule. Thankfully, the edge of a wood or a nice thick hedgerow is much more their cup of tea.

Out in the country, they rarely need to be moved and some setts, as I have indicated earlier, are used and re-used over hundreds of years. The problem ones are those that create dangers, which undermine properties or are simply in the way of a new development. In all my years of studying badgers, I have come across some pretty odd locations, and more recently in my role as consultant I have had to make safe or remove setts from some potentially hazardous sites. Let me mention just a few and as I do, I shall indicate, very briefly, some of the solutions.

Let's start with canals. Banks built to retain millions of gallons of water are not improved by nature's version of a JCB setting up home in a series of self-dug tunnels and chambers that weaken the structure and create the chance of a serious loss of water. So, when badgers decide to establish their setts in elevated sections of canals, this creates very serious problems for British Waterways engineers. There are well-documented instances of this happening for the past 200 years or so and, before badger sett protection came in, it was the policy, both with badgers and foxes, to evict them by throwing a bucketful of creosote into the tunnel system. Both animals have an acute sense of smell and as neither would be prepared to put up with the over-powering smell of creosote, it would drive them out. That is no longer legal, so alternative solutions are now needed. In my experience, certain lengths of canal are more vulnerable to badger damage than others and most problems occur in low-lying areas where the water table is high, and where, if badgers were to dig a sett, it would fill with water. The canal banks, by contrast, are elevated and well drained and so it is scarcely surprising that the badgers choose them to set up home. The same applies to railways and roads.

Since 1991 I have worked in many different locations on canal bank problems and the solutions have followed the same broad principles. Short-term measures simply delay the inevitable – concentrate on the tiny area containing the badgers and they will soon be back. The remedial work has to be much more extensive. My approach entails removing approximately 25 cm of topsoil and vegetation, securing weld-mesh to the embankment for a distance of as much as 50 m and then replacing the topsoil.

The vegetation soon recovers and, when the badgers return, as they inevitably do and attempt to excavate a new sett, they come to a halt when they come up against the weld-mesh. Even they can't cope with that kind of armoured plating, and they give up. The mesh also renders the bank rabbit proof.

So much for canals; what about roads? They are a killing ground for wildlife, no matter what precautions are taken. New roads are a particular problem, as they are often built across centuries-old badger tracks, cutting badgers off from long-established feeding grounds. In those circumstances the badger and other ground-feeding animals, either have to be stopped by some impenetrable barrier or have to be helped or encouraged to get to the other side in safety. If a fence is put up, it has to be a substantial one, for badgers are accomplished climbers and as we know, are also inveterate tunnellers. I have witnessed badgers climbing to a height of several metres; they have the strength to get through flimsy fencing and will tunnel underneath ordinary fencing unless prevented from doing do. So badger-exclusion fencing along a roadway needs to be of solid, durable construction, and the Highways Agency has a tried and tested design that works well. Fences can be used, either to stop badgers in their tracks and turn them back, or, more usually, to divert them so that they go over or under the road. Happily, they quickly adapt. I recall seeing a new bridge being built to carry a country lane over a new road. Badgers used it the very first night, crossing over it as though it had always been there. On another occasion, earth-moving equipment had taken a huge amount of soil, the size of a dual carriageway, from a field. Later that evening, when the contractors had gone, a badger crossed through the excavated section, following the line of what had been its track. It kept going as though nothing had changed, reached the other side of the workings and trotted casually on down the track.

Culverts underneath existing roads, bridges and railway tracks have long been used as safe, ready-made crossing points for wildlife of all sorts. When new roads are being built, underpasses can be built into the design, and these should be placed on existing main badger tracks. Badgers quickly accept them, though if they are too long or too complex there is a danger that they may be used as setts, or alternatively that foxes will take them over as earths. I know of several sites where this has happened. It is my experience with new roadworks that, if a usable culvert exists within 100 m of where the new road crosses an established badger path, then the badgers will probably prefer to use these rather than the new ones.

Despite all the noise, vibration and the hazards of living so close to traffic, badgers do often build setts under roads and I have been called in many times by councils to advise when badgers have undermined roads. The first incident I dealt with was in 1991. When I arrived, I found that half of a B-class road had caved in and a one-way traffic control system was already in operation. The cause of the collapse was obvious enough – badgers had been tunnelling underneath. On the field side of the hedgerow, I found a 28-entrance badger sett with 22 of the tunnels leading directly under the road. The spoil heaps consisted of large stones and tarmac and I was told that, over the years, council workers had filled in surface holes as and when they appeared. It was obvious that the badgers had excavated much of that fill-in material from below. I decided that to evict the badgers would not be a lasting solution as they would simply move further along the road and cause the same problem. It would also be less stressful to them to carry out remedial work with them still in the sett, so the road surface was removed under my supervision (remember, I take full responsibility for the badgers' safety, as should anyone providing a similar service). This exposed a large number of badger tunnels, and investigation showed that some of the tunnels had as many as five layers of tarmac where the road surfacing had collapsed into them over the years. All the exposed tunnels were carefully reinstated using 30 cm diameter pipe. Reinforced concrete was laid over the top and the road was then re-surfaced in the normal way. It took two days to complete the work and in all this time, no badgers were seen, either in the sett or leaving it. Since then, I have revisited the site many times, the last time at the end of 2008. The sett has increased in size, but there have been no more road collapses. The council are happy and so are the badgers. I have used similar renovation methods many times since and all have been successful.

Roads create one set of problems, but railways are quite a different proposition and it is rarely possible to deal with them in the same way. Most often, I use the same technique as I do with canals, evicting the badgers, having first carried out a survey of the surrounding area to satisfy myself that the badgers are not made homeless and that there are setts close by that can be used. One especially tricky situation comes to mind. I was called in to look at a main line intercity track, having been warned by engineers that the subsidence was getting towards danger level and that the badgers had to be evicted immediately. Unfortunately, it was March, and I recognised from visual signs around the sett entrances that there were cubs underground. I told the engineers that I could get the adults out but not the cubs, because they were too young and not ready to move. So what could be done? I suggested that, as the track was monitored weekly for safety, I could monitor the sett with the same frequency. As long as the badgers did no more tunnelling, we might get by until July when, under licence, we could evict both adults and cubs. This is what we did and the badgers promptly enlarged a small secondary sett in a hedgerow close by. I recommended that the entire embankment, a length of 50 m, should be sealed off with weld-mesh to prevent the badgers re-excavating the sett at a later date. Unfortunately, my advice was not taken and when I visited the site two years later, the badgers had returned, just as I had warned. Thankfully, Network Rail now have a much more positive strategy and really do care for all wildlife and the safety of the track.

Housing developments, small and large, frequently need the involvement of a badger specialist and you don't need me to tell you that any proposals that will affect wildlife, even in the short term, often create all sorts of strong emotions – mostly from objectors. Most amateur naturalists or keen conservationists are horrified when a site is earmarked for development; they fear for the wildlife and think that it will be lost to the area forever. Experience shows this is not the case. House building brings disturbance, of course, but once the new houses are established and lived in, all sorts of plants, bushes, trees and shrubs start to grow. Green verges and well-kept lawns replace what, in many instances, was unproductive or unkempt rubbish-strewn scrub or low-grade, almost sterile, arable land. Provided that planners, councils, developers and the like all play their part, every housing development is also preceded by a phase 1 habitat survey. I am called in when badger activity is evident or suspected and I make it my first job to survey the entire area, including all the boundaries and what lies beyond. On the site itself, every square metre is examined. There is no easy way to check out a derelict, overgrown area and it often involves crawling on hands and knees through clawing undergrowth that scratches the skin, tears at clothes and starts the heart racing with the effort. But it has to be done – miss one clump of uninviting bramble and you can be sure that is the one containing a badger sett.

Those, then, are the usual locations to which the badger consultant is called. But I'd like to add one more example: airfields. In my home county of Shropshire, I was called out to an airfield at Shawbury, the military helicopter-training centre where Princes William and Harry completed their training. I was called in to deal with an occupied subsidiary sett just 40 m from the main runway and only 10 m from a helicopter landing area. Close by was a large concrete area where Harrier jump jets were anchored down and their engines tested – the noise and heat was beyond anything that I had ever encountered. The resident badgers were subjected to extremely high noise and vibration levels. But, whose fault was that? They chose the site and seemed quite content, despite all the activity. Indeed, it proved extremely difficult to discourage them from staying there. Not that they were alone; other wildlife seen at night on the airfield included deer, foxes, rabbits and polecats.

Just one more thought about new roads: they aren't always bad news for wildlife. Some years ago, construction of a new road was proposed close to where I live. It was scheduled to go right through my badger study area and the local chairman of the CPRE (Campaign to Protect Rural England) asked me on what grounds they should object. To her surprise I said they should welcome it with open arms. As the proposed route crossed arable land, I took the view that there would be far better habitat after construction than there was before; there would be more permanent grassed areas and a variety of small trees and shrubs would be planted. It is worth bearing in mind that, although we always seem to regard the badger as a country animal, it is an adaptable creature that has learnt to live alongside humans. A large proportion of my work is in cities and towns, among them London, Birmingham, Coventry and Manchester. The badgers, no doubt, were there first, then the developers arrived, up went the houses and office blocks. The badgers made the best of what they were left with and stayed there with development continuing to expand around them.

Sett closures and artificial setts

When I am giving talks, I'm often asked about artificial setts. When is it necessary to build one? Are they always the only alternative if a sett has to be closed? And if they have to be built, what size and design do I recommend? Let's deal with that second question first. In open countryside, if an existing sett has to be closed, it is unlikely that an artificial sett will be necessary, the badgers will simply move into another of their smaller setts, a subsidiary or an outlier and enlarge it; this doesn't take them long if the ground conditions are generally favourable. But, if the sett to be closed is a main breeding sett in an area where the badgers would have few if any alternatives, then an artificial sett may well be the only satisfactory option. This happens quite frequently where badgers are living close to a built-up area or housing development of some kind and where the remaining parcel of open land is scheduled for development. On the edge of a town or city, for example, you may already have buildings on three sides and if the new development would enclose the badgers or cut them off from their foraging, then the only option is to build a replacement sett close to good foraging.

How big should an artificial sett be? To answer that, you have to look first at the sett it is replacing; if a main sett is being closed, then its replacement also has to be quite large. Main setts are often quite extensive, so the new sett should also be roomy. I closed one main sett, which was found to have 22 chambers, so the artificial sett I designed to replace it covered 20 m by 30 m and had

12 chambers. Not as many as the one it replaced, but the badgers took to their new home very readily and subsequently enlarged it. Since then, I have built many more similar-sized setts that badgers have moved into and bred successfully.

The first thing to do in choosing a site for an artificial sett is to determine the water table – that is, the height to which water in the ground will rise during the wettest part of the year. For the site to be suitable, there needs to be at least 1.5–2 m of soil above that level. If there is less than that and there is no other satisfactory alternative site, the remedy is to build on top of the ground and then cover it with at least 1.5 m of topsoil. On development sites, large quantities of soil are generally readily available.

Badger social groups usually have a number of outlier setts in their area and they can sustain the loss of a couple of these without any hardship, so normally there is no need to replace these with an artificial substitute. Always bear in mind that the sett is not the be-all and end-all for a social group of badgers: it is their territory that is really important. If there is an ample supply of food, they will tolerate a tremendous amount of upheaval, but if they cannot find enough food, they will move on, no matter how large the sett. If three or more outliers or, indeed, a subsidiary plus a couple of outliers are taken out to make way for the development, then normally, in my opinion, a replacement sett would be required. A design incorporating four to six chambers and covering an area of 20 m by 20 m would be adequate. Ideally, it should be built following the design principles I am about to explain.

If it has to be built above ground and be covered with soil, then I think the minimum size should be 20 m by 20 m. The design of an artificial sett is critical to the long-term success of the project. Just as we like our comfy armchair in front of a nice fire or in a warm room, so badgers too enjoy their animal comforts. To my dismay, however, they are still expected by some consultants and some badger groups to put up with second-rate, unsuitable accommodation: cold, dusty chambers made from concrete, with tunnels constructed from concrete pipes or, even worse, glazed pipes.

How on earth can an animal be expected to get a good footing in a glazed pipe or on a round concrete pipe? I know they will use them from time to time, but only for short periods. My advice is to make them as natural as possible. I like to feel that the setts I build replicate as far as possible what the badger builds for itself. The tunnels and entrance holes reflect the body size of the animal. In practice, that means that the piping used needs to be no larger than 30 cm; 40 cm is too large. I don't use concrete. I much prefer to use twin-walled plastic drainage pipes with 10 cm cut off the bottom. In profile, the piping as it is laid looks almost horseshoe shaped. This allows the badger to walk on a flat surface and, more importantly, on earth. Badgers take to this type of tunnel very quickly, the soil underneath is natural and smells natural – very different from cold, slippery, inert concrete, which is a poor choice.

Plastic piping also has other important advantages. It can be easily cut to any length, and holes can be cut into the sides to allow the badgers to tunnel out in whatever direction they choose, to increase the size of the sett. Plastic is so much better than concrete and glazed drainage pipes that I think there should be regulations requiring its use. I know that badgers and foxes will choose concrete culverts as homes, but if you look inside these you will find all sorts of debris that has accumulated over the years, so the occupants are actually walking on a flat, almost natural, surface very similar to the design I recommend.

What about badger chambers? In books, I have seen designs that variously recommend the use of concrete, brick or breeze blocks. I have even seen designs with a damp course membrane and gravel or even wood chippings on the floor. This does not create the right environment. Tunnels and chambers should not be dry and dusty. The dust will irritate badgers' nostrils and that is not what we want happening to an animal that relies so heavily on scent to feed and navigate. Every sett I have opened has had a damp floor. The air throughout badger setts is invariably damp; it has to be to enable the badger to utilise its fine sense of smell. In nature, there is no dust in a badger sett and no unnatural dryness. The badger's underground environment is almost completely dust free and even the bedding material is invariably slightly damp. So, please, to anyone contemplating making an artificial sett: no more concrete floors, no more brick and concrete sides. All you need for a chamber is shuttering-grade plywood untreated and not weatherproofed. Add to this four wooden stakes, again untreated. You do not need nails or glue to make the chambers. Simply create a chamber by driving the posts into the ground to form a square, take the pieces of plywood (about 60 cm by 60 cm), with holes already cut into them where the pipes will join on and tap them into the ground, using the posts as supports to prop them up. Place the soil on the outside of the ply, thus keeping the boards in position, and then place another piece of ply on top. We want that chamber to be damp, remember and each chamber should have two entrance or exit holes. In a multi-chamber sett, not all of the holes need have pipe tunnels leading into them: some can simply be left as holes through which the badgers will excavate their own tunnels at some later stage.

So much for the basic structure. Building the home is one thing; getting badgers to move in is another. But we can do things to help. I have already emphasised how important scent and smell are to the badger and just as we recognise our homes by sight, badgers recognise theirs primarily by scent. To take the house-moving analogy a stage further: when we move house we take with us the things we cherish and which are familiar to us, so to adopt that principle when badgers are being moved, we simply need to transfer some of the scents that are so important to them. The first thing I do when I am building a sett is to take some soil from the spoil heap around the sett that is being closed and which is impregnated with the badgers' scent. Then, as I build the sett, some of this spoil is trickled along the floor of the tunnel system. When the sett has been built, I then put more of the spoil around the entrance. In addition, I take dung from the badgers' latrines and smear it near the entrances too. That way, the new sett has many of the familiar scents of the old one. More often than not, this is all the encouragement they require, but as a bit of extra insurance I lay a trail of peanuts from their existing sett towards the new one. First, I sprinkle a few close to where they are living, then the next night I put down more, including some a little further away and outside the replacement sett. I repeat this night after night, laying down a small handful of peanuts every 10 m or so to keep them heading closer and closer to where they will ultimately live. Ideally, the new sett will be in a favoured part of their foraging area, which means there will be a well-used badger track leading from the old sett to the new one.

It is important to do this preparatory work regularly and it is best to put the peanuts down late in the evening when most of the birds have finished feeding. Putting peanuts down inevitably attracts attention and whenever I do this I am aware I have an audience (notably magpies, crows and squirrels) watching my every move. As soon as I move away, they move in, anxious for a free banquet. To prevent this happening, I first started to place small pieces of plywood about 30 cm by 30 cm over the nuts, but soon found that the crows would fling them aside and then tuck into the nuts. I solved the problem by placing a large stone as big as a brick on top of the ply. This proved too much for the birds, but the badgers had no problem removing it. Whatever else you do, do not hang around to see if the badgers are finding and taking the food. Leave them to it and do not watch them because they will pick up the human scent and it will then take longer for them to settle into the new sett.

If the badgers haven't ventured inside the new sett within a night or two, then you're the wrong person for the job! They won't necessarily live in the sett as quickly as that, but they will visit every night. I find that badgers usually visit a new artificial sett on the first night it is built and badger paw prints are found at, or in, the entrances. This has been the case with every sett I have built. It must be strange for them because it is impregnated with their own scents, which, of course, they didn't put there. The sett will interest them, they'll wander in and out and with each visit they will deposit more scent. Reassured, and gathering confidence, they will soon accept the sett as their own.

How can you be sure they are actually living in the sett? Well, you can sit there and watch for hours on end, which, as I say, I don't recommend, or you can do a bit of detective work. If the soil at the sett entrance is nice and loose, simply rake it over each evening and check in the morning for paw prints. If it is stony or too hard packed, sprinkle a layer of sand at the entrance, then the distinctive prints of the badger will show very clearly. Additionally, you can place small twigs just inside the entrance each evening and check the following day to see if they have been flattened. From prints you can tell whether badgers, foxes or rabbits have been active at the entrance. In really cold frosty weather there's another sign of habitation to look for: steam drifting from the entrances is a sure sign that the sett is occupied.

Is it worth putting bedding in the chambers? I don't think it matters too much either way. If I'm building a sett with around eight chambers, I do put bedding (hay, preferably, but straw will do) in two of them and leave the others empty. Always have the replacement sett completed and badgers using it, before attempting to exclude the badgers from the sett that is being closed. Natural England certainly would not allow you to close the sett until the badgers have started to visit the artificial sett. That is as it should be. Ideally, the artificial sett ought to be ready for occupation at least four weeks before the natural sett is closed. If the artificial sett is built too soon, it is likely to be taken over by foxes.

How close to the sett being closed should the artificial one be? In part, this is a matter of judgement, a test of the consultant's expertise – the signs have to be read correctly. They must be confident about the badgers' existing range and certain about how far they are travelling to their preferred foraging grounds. Provided they have worked that out correctly, the new sett should be sited in what appears to be the most suitable available spot, taking into account factors like natural drainage, natural cover, the amount of noise or interference likely and any restrictions that the developers themselves will impose. Terminology is interesting here. Many people talk about the badger's 'territory'. To me that is land they will defend vigorously against other badgers that are not from their own family group. 'Range' I think is more apt, because it indicates the ground over which badgers travel to find their food and it is my view that the ranges of several different social groups will often overlap, the borders between them, in other words, not being absolute.

Safeguarding the new sett from attack is another factor that has to be considered. In areas where badger digging is known to occur, or on a site where the badgers look to be potentially vulnerable, some sort of added security is worth considering. Steel mesh placed over the sett, extending beyond it and covered with a good depth of soil is one standard technique used to good effect where digging and baiting is still practised, though these days new badger-capture techniques are being used, which are less laborious and unfortunately just as successful. The baiters wait until the badgers have left the sett, place nets over the entrance holes and then disturb them where they are foraging. The badgers rush back to the setts, where they become entangled in the nets. The baiters have got their badgers and within minutes the scene is quiet.

Once the badgers have moved into the artificial sett, give them plenty of time before thinking about doing any badger-watching there. I would recommend giving them at least six months to settle in. By all means, visit the sett occasionally in daylight to check that it is still in use, but otherwise leave them alone. I am against large numbers of people sitting close to any sett; two or three at a time is plenty in my view and with a new artificial sett, I recommend just one or two in the early stages.

Timing is important in artificial sett construction. Under Natural England's existing rules, badgers can be excluded only between the 1st of July and the 30th of November, although I have suggested to Natural England that they reconsider these timescales. My experience tells me that, as winter nears, badgers become lethargic as, after rich autumn feeding for several months, they prepare for reduced activity; so, with that factor in mind, I would rather that badgers could be excluded only until the end of October. I also believe that, with the exception of main setts, exclusion could start earlier in the year – at the beginning of June. This would be less stressful to the badgers than November. From this, you will gather that the best time, in my opinion, for badger exclusions to take place is from June to October. Construction of the artificial sett ought to be completed 4–6 weeks before sett closure. Just to remind you about the badger breeding cycle. Cubs are most commonly born around mid-February, and usually appear above ground in mid-April, but there are plenty of exceptions to this rule. The earliest I ever saw cubs above ground was the 14th of February. I remember the occasion well, for in my excitement I also forgot it was St Valentine's Day and I didn't take my wife out for a meal!

I would like to indicate one problem with artificial setts: it seems that there is a greater risk of foxes moving in and taking over these than there is with longer-established (and probably larger) main setts. Badgers will tolerate foxes taking over part of a large natural sett, but these unwanted, uninvited squatters are not ideal when you are trying to get badgers to move into a totally new sett. Foxes are no match for badgers if it comes to a trial of strength of course, but you simply don't want the problem to arise.

Just a few words here about the mechanics of badger sett closures. Let us assume it comes under the jurisdiction of Natural England, rather than DEFRA. Natural England issues a licence and imposes a number of conditions that must be met in excluding the badgers. These determine, among other things, how close the contractors can come with earth-moving machinery and other heavy equipment, and they may also indicate the method of exclusion. Natural England used to stipulate that two-way gates should be placed over the entrance holes for a period of seven days or so, to allow badgers to continue to move in and out of the sett during this time. They then said that these should be replaced by one-way gates for a period of at least two weeks. I very much disagreed with this approach: the decision to exclude had already been made and alternative accommodation, whether natural or artificial, was already available. Day 1 should be the decisive day. Once the gates are in place, you don't want the badgers to return. Happily they no longer recommend two-way gates. At best, it is unsettling for the animals, but at worst it is not far short of mental cruelty – the animal cannot work out what is happening. Once the badgers have been excluded and before the sett has been demolished, there is the danger that they will return and try to work their way back into the sett. Often, they try to dig around the side, top or underneath, so to deter them I place a large piece of plywood about a metre square on the entrance floor immediately in front of the gates and some at the sides; that prevents them gaining access.

While the gates are in position, it is important to establish whether any badgers remain in the sett. To do that, I place three pencil-sized sticks in the floor of the tunnel, at roughly arm's length beyond the gates. I also put a twig on the outside of the gate. If badgers are inside the sett and merely come as far as the gate, the twigs inside will be flattened. If badgers exit the sett, the single stick outside the gate will be displaced. Surveillance and close inspection of the area around the entrance will help to determine whether there has been any vandalism or visits from other wildlife that may have caused that stick to be moved. My preference is to check the sett every day until the sett can be closed. So far, the longest I have had to wait for that to happen is six days and normally they are out within two days.

Let me tell you about the one that took much longer, for it taught me a great deal. I was asked to close a six-entrance sett in the dam of a small, nine-hectare lake sited above a village. The sett was a safety hazard, so it had to go. Not too far away were two other large secondary setts, so an artificial replacement sett was not needed. I soon had two of the badgers excluded, but I knew from the way my indicator sticks inside were being moved that there was at least one badger still in the sett. This went on for six days. On the seventh day, I decided to walk along the top of the dam and I found a fresh hole, excavated from the inside. Inside it, I could see the torn and severed roots of vegetation, including the roots of a small tree nearby. I pushed in a steel tape measure. The hole, which was dug out of solid impacted clay, was vertical. It measured 97 cm deep and 15 cm across. The badger had decided not to come out of the gates; he'd chosen instead to dig his way out. It was quite an eye-opener and a remarkable demonstration of power. To remedy the situation, I placed a one-way gate on the escape hole and the problem was solved.

A few words of warning for anyone likely to be drawn into badger exclusion work – to be entirely satisfied that a sett is empty and disused, I make it my practice to crouch down and peer inside a sett. I look for cobwebs and for signs that tell me the sides of the tunnels are drying out and starting to flake – rather like emulsion paint peeling off. Try that for yourself, but take care, for you are likely to attract some uninvited boarders. Later, walking home or driving back in the car, you may begin to feel itchy and fidgety. A few bites in some delicate places will soon tell you that you have some unwanted company – badger fleas! The chances are that you are the only warm-bodied host they have encountered in several days, so they have taken the chance to jump on to you. Their opportunity is your infestation problem! I can assure you from bitter experience that they are the devil of a job to get rid of.

We were once called to carry out a badger survey for a quarry extension in South Wales where a large badger sett on the edge of the working quarry needed to be closed under licence. A great deal of time was spent looking at all the options, then a large artificial sett was built to accommodate the displaced badgers, but we found that a second sett close to this one could be saved with a protection zone around it. After many weeks of monitoring the artificial sett and when we had evidence that the badgers were using it on a nightly basis, we decided to close the sett, which was then demolished by a very large excavator. The bank containing the badger sett measured 14 m deep from field level to the quarry floor and as we cautiously went deeper into the sett, we found that it was on three levels with 15 chambers, one of which was gigantic. The machine driver was very experienced and had adhered to our request for caution throughout the excavation. We progressed towards the large chamber a few centimetres at a time and when half the chamber had been exposed, it measured 1.2 m in height and 1 m in width, with five tunnel entrances in the bottom and a considerable amount of bedding. This chamber indicated that we were dealing with a sett that had been in occupation for probably 100 years or more. The size of this chamber would be caused by roof collapses over the years – the badgers would have removed the collapsed soil and deposited it on the spoil heaps.

Whilst we were monitoring the artificial sett, my assistant and I sat to observe badger activity in the immediate area. All was still, the light was fading fast and visibility with the naked eye was poor, when a bird with a white front landed and was walking in a searching manner in the grass of the pasture field. At first, I thought it was a magpie and then questioned why a magpie would be out at this time of night. When ten minutes later what we recognised as a barn owl swooped in and landed close to the first bird, using our binoculars we realised that the bird feeding on the ground was also a barn owl. Both owls searched the ground and were feeding for more than 20 minutes – they were obviously eating invertebrates, possibly earthworms. We found this very interesting and it explained something that had puzzled me for over 30 years. I had on previous occasions found the remains of barn owls that had been eaten by foxes and I wondered how a fox could possibly have caught a barn owl as, when they are hunting, they normally just swoop in and grab their prey without landing. I now understand that, whilst their minds are occupied feeding on the ground, a fox can catch them unawares.

Tackling the big jobs

So much for closing setts of fairly modest size, but what about those with 20 to 30 entrances? What do you do – gates at each and every entrance? No, that isn't practical. The answer is an electric fence, operated in conjunction with a small number of one-way gates. However many entrance holes there are, some will be cleaner, more highly polished than others and it is probable that these will have well- defined tracks leading to them. Either side of the gates you need steel electric wire fencing, but not flexible netting electric fence. Flexi-fencing is, in my experience, inefficient: too many things to go wrong. My preference is to install four strands of electric wire, fixed to multi-strand posts. You can either choose to have all four strands live, or, and this is the method I prefer, have the bottom strand as earth (it doesn't matter if the grass touches it) and low enough so that, if a badger puts its head between the first and second strands, it will touch either one. Because you have a good earth, the electric fence is more efficient. It will not injure the badger, but it will give it a nasty surprise. With the four strands in position (three of them live), one-way exclusion gates are then placed on the badger tracks and the electrified wires are looped over the top of the gates, because the badgers may try to scramble over the top at some later stage to try to get back in. Using this system, I have had 100% eviction success of badgers from very large setts, even when others before me have tried and failed. Electric fencing must have a message attached to warn the public. This sort of wording will do: "*WARNING. This fence is electrified for animal control. KEEP OUT!*"

Sand sprinkled on the ground each side of the gates and a twig set against the outside of the gate will tell you whether badgers have passed through. If the grass is very long around the sett, a metre-wide strip should be cut around it to accommodate the electric fencing. Check each entrance every day – if the badgers are used to human presence, this should not be a problem. But, if the sett is at a location where there is usually little or no human activity, they will pick up your scent. This often happens on agricultural land when you are doing work under DEFRA licences, so you have to bear in mind that this will deter the badgers and the exclusion may take longer. Always exercise caution, whatever the location. The fewer disturbances, the less the stress to the badgers and the more likely the exclusion is to work quickly.

I remember being called out to a sett where ecologists had been called in to carry out the exclusion. It had not worked for them and I soon found out why. The badgers were leaving the sett via the one-way gates, but were getting back in by climbing over the fence put up to keep them out. When I was called in, work to excavate the sett had already started and the JCB used had begun from a position on top of the sett, the method excavator drivers seem to prefer. I do not like that as a method, nor do I like JCBs (or earth-movers of a similar design) for this type of work. When they put their jacks down to stabilise the vehicle, there is a danger that these may penetrate the tunnel or chamber systems. I know of at least one instance (not one of my jobs, by the way) in which a badger was badly injured and had to be put to sleep due to this.

Closing down a sett has to be done methodically, professionally and with a great deal of care. My choice of earthmover is a wide, track-laying machine. I don't allow it to work from the top of the sett because, despite all the exclusion precautions that have been taken, there has to be some slight risk that an animal, not necessarily a badger, may be inside. The earth-moving vehicle needs to approach the sett from the front, and should excavate a trench at least 1 m deep, some distance from the spoil heaps, because invariably there are sleeping chambers under the spoil. The machine must then work carefully into the sett under guidance. This is the stage where the consultant, if he knows his business, takes over detailed control, directing the machine operator and guiding him every inch of the way. My preference is to start 4–5 m from the sett entrance, work systematically towards the sett and through the tunnel system, following each tunnel in sequence to its end. At all times, you should be able to see half a metre or a metre ahead of the machine, using a torch if necessary to satisfy yourself that the tunnel is empty. It should be, if you have done all your preparation correctly, but there is no point in taking chances; no animal deserves to suffer. Please note that an animal in a sett is classed as a captive animal, and it is illegal to injure any captive animal, including foxes, rabbits and rats. It is probably enough to take out 30 cm slices of earth at a time from around the tunnel. Not much, I know but, if the consultant and operator work together carefully, it is surprising how quickly the job can be done. As you near the end of a tunnel, it is worth an extra careful look. If a rabbit has taken refuge there, simply stand back and let it bolt for cover, then carry on. That has happened to me several times in the past.

Not many of us get the chance to see inside setts in such detail, so I make a point of having someone at hand to help, who can make a note of the tunnel lengths as I call them out. I also indicate the number and approximate size of the chambers. Together, we try to map out each twist and turn of the network of tunnels as we watch the earthmover at work. Do not expect to find many chambers with bedding or to find dust. In chalky soil, you get fine, powdery deposits, but not dust as such. Some chambers will have old, rotting bedding with some mould on the top; other bedding will have some mould on, and that is potentially hazardous for anyone who suffers from asthma, as I do. The chambers with dry, but not dusty, bedding are the ones to be careful of, as these are the ones with fleas, those would-be lodgers looking for a home, for someone they can literally latch on to.

I would advise anyone involved in sett closures to start early, take food along with them and if at all possible to work on with only the shortest break. It is very important to complete the closure in one day. That saves having to put the electric fencing around what is left and ensures that, in any event, nothing is left for the badgers to return to. My records show that setts with as many as 200 m of tunnels and 30 or so chambers have been closed in a single day depending on the soil strata and the depth of tunnels. One sett was three-tiered and dug to a depth of 14 m, but we still completed the closure in a day.

A word or two about badger gates. There are lots of different designs – solid, weld-mesh, wooden, steel-plated and so on. My preference and the design I have always used, is weld-mesh with a gate size of about 40 cm by 30 cm. The weld-mesh has a 4 mm gap and the distance between each structure bar is 75 mm. This is very close mesh and despite what you might hear from others who say badgers will force open such gates, I have never yet had a badger that has managed to open them to get back through: in through the side, yes (though I now prevent that happening), but not through the gate. I place them at a slight angle and, though the badgers may return to the sett and claw at the gates, they fail to re-open them. Why? Well, what people forget is that it is not physically possible for any animal to lift its paw above its head and walk forward at the same time. It can do one or the other, but not both at the same time. That makes it impossible for a badger to pull the gate open. Why do I so favour weld-mesh? Well, just think of what happens. With a solid gate fixed firmly onto sett entrances, you are reducing air flow almost totally, and that is not a good thing. Wooden gates are a definite no – they warp and will become stuck in an open or partially open position, and so they are simply unreliable. Weld-mesh allows free air movement and badgers in the sett can actually scent what is outside. In my view, it is much more natural and badger-friendly than all the alternatives.

Chapter Seven

Badgers and farming

Too much muscle

Thousands of years ago, before farmers created the pasture-lands that are today's rich source of earthworms, badgers were primarily scavengers and meat eaters. Their immensely powerful jaws are a reminder of the days when they needed the strength to crush bones and tear carcasses apart, much as hyenas still do. Today, although these powerful jaws come in useful for shearing off tree and hedge roots encountered when they are excavating their tunnel systems, badgers, surrounded as they are by rich farming lands and woodlands, are quite content for most of the year to feed mainly on worms, insects, berries and fruits.

Yet, I have lost count of the number of times farmers have rung me in mild spells early in the year to complain, "The badgers have been at my lambs again. They have had two or three in the night."

"Are you sure it's not foxes?" I'll ask. "No, it's badgers; their hairs are all over them." They'll go on to describe what they have found: dead lambs, partly eaten by badgers. OK, something, most probably foxes or badgers, had been feeding on them. But, did they kill the lambs, or did they simply come across a carcass? I have no scientific evidence for saying so, but my instinct and experience tell me that, when badgers are involved, it is most often as scavengers, not as killers. The lambs have been born dead, or have died shortly after birth, perhaps from diseases like *E. coli*, which is the cause of many lamb deaths. The carcasses have remained unburied and the badgers have then fed on them. Only a very hungry badger would kill a live lamb. It is significant I think, that, as yet, I have never met a farmer who has actually seen a badger kill a live lamb, though, having said that, very few have seen a fox go for the kill either. My guess, backed up by surveys produced by MAFF (now DEFRA), is that at least 95% of lamb deaths are due to natural causes including, of course, disease. Many fall victim to *E. coli*, as do young animals of most species, something that as a former farmer I know from bitter experience. *E. coli* weakens the animals, and antibiotics have little or no curative effect; the lambs can seem in the best of health one moment, but be dead within a few hours.

Those who prefer to blame the badger or fox and who quite illegally shoot badgers (sadly now happening on a large scale following the bovine TB controversy), should stop, think and use the evidence of their own eyes. Time and again, the evidence is there if they are only prepared to look for it. Anyone who knows animals – a vet, for example, or a good stockman – and who is prepared to take an objective view, can look at what remains of a carcass and nine times out of ten will be able to tell whether it was a fit, strong lamb that died. Most of the lambs I have examined that were allegedly killed by foxes or badgers, were weak and in poor physical condition, with scour stains around the tail.

However, not all stock losses are the result of *E. coli*, as I learned the hard way when pig farming. Years ago, as far back as the 1940s, pig farmers were paid on the basis of length of carcass. So, what did we do? We bred from selected stock to ensure we earned the highest price. But, breeding for length weakened their backs and sows would have problems with their backs at 2–3 years old. A long carcass also meant a higher fat-to-lean ratio. Then, the market began to demand leaner meat and breeders began to select stock that gave them that; the new generation pigs were thicker, broader and more muscular. I earned the reputation for lean pigs of exceptional quality; up to 93% of mine were graded as 'double Qs', which was the very best quality. I had selected for lean meat and also had a quick generation turnover. Through selective breeding, I have seen changes in my lifetime that would take thousands of years with natural selection. But, after a few years of successfully rearing this new leaner pig, I started to find that some – always the largest and strongest of the litter – were dropping dead in their pens; there was no disease, no wasting and to look at them they appeared to be perfectly healthy. Perhaps 0.5% or fewer of my pigs were lost that way and of course, I wanted to know why, so I called in the vet. He looked at the hearts and found that, in the process of rearing pigs to give virtually fat-free meat, we had put extra muscle, not just on the carcass, but on the heart too, externally and internally. The extra muscle had closed in on the heart chambers, which had become too small to cope with the blood supply needed. The result: sudden heart failure.

What's all that got to do with badgers? Well, some years ago a farmer for whom I have the highest regard (his stockmanship is second to none and he is also keen on wildlife conservation) rang me with a problem. "George", he said, "I tolerated losing a couple of my best lambs, but now I've lost another. They're all about three to four weeks old and in their prime, beautiful animals, the best lambs in the flock and I know the badgers have killed them. There's badger hair all over them. What can I do? I'm told that I'm in my rights to shoot badgers if I see them attacking my flock."

"That's right," I said. "If you actually see them attacking the flock, you can legally shoot them. But once you know they have been attacking your sheep, to shoot them is then illegal. In other words, now you know, you will in fact be committing an offence."

"All right, then," he said. "What can I do?" I told him to put an electric fence around the field. "Better still", I said, "Surely you have another field, further away, that you can put your lambs in. That way you'll have a better chance of establishing the cause of death."

"Right, I'll try that," he said.

"See you tonight," I told him. "I'd like to have a look at this latest lamb you've lost."

I called round and sure enough it was a superb animal, in excellent condition and there was badger hair on it, especially on the ribs that had been chewed.

Then I turned to him. "You're one of the top sheep breeders. Tell me, what are your quality grading results?"

"Pretty good," he said. "They're about 90%."

"Thank you," I told him. "That's what I wanted to know. What's happened is that the lamb probably died of heart failure and the badger has eaten it."

And then I told what had happened to me all those years ago.

He looked at me, shocked. "Good, God," he said. "I think you're right. They are my best lambs, not small or average. It sounds like the same problem."

Events subsequently proved that it was. The big improvement in lamb quality – less fat, more lean – had come some years behind that achieved by pig breeders, but the result was the same. Prime lambs were dropping dead. Badgers and foxes were getting the blame when, in fact, all they had done was to scavenge. Moral: it is too easy for farmers to jump to conclusions on the basis of prejudice rather than fact.

To repeat what I said earlier: I have yet to find a reliable witness who has seen a badger attack a live lamb and I have yet to see it for myself, despite all my years in farming and all the countless hours I have spent badger-watching. What I *have* seen, often, are badgers and foxes foraging in fields alongside sheep that have taken not the slightest bit of notice. By contrast, if a dog enters a field, the sheep will flock to the furthest point in the field. Once and only once, I saw a badger make a dash for a hen. The badger was one I was rehabilitating and the hen was in my yard, and did that badger move! It absolutely flew across the yard. There were feathers everywhere, but somehow the hen survived. The badger had blundered in head-first. A fox would have crept along the side of the wall, sneaked closer and then it would have scored a kill. I am not, however, saying that badgers *never* kill farm livestock, as you will see in the next section.

"Killed a hundred, he 'as"

Even the occasional act of predation by badgers provides ammunition, of course, for those who take exception to any of our wild creatures impinging on their way of life. Consider the case of the elderly farming couple who contacted the RSPCA and complained that badgers had killed all their poultry. The RSPCA called me in and asked if I would check it out. I said I would and set off, convinced that foxes rather than badgers would be to blame. I arrived at the farm and found that most of the buildings were dilapidated and gates were hanging askew off their hinges. I was met by the elderly couple and the old gentleman took me quickly to the hen pen in the stack yard. Looking round, I saw badger prints in the farmyard mud that was sheltered from the frost. We had had a few nights of frost and badgers and foxes are more likely to attack farm livestock during a frost, as their usual food supply (earthworms and grubs) is sealed off underground.

"Come on", he said, ushering me over to the pen. "Here's the chickens. This is what the devil has done." We reached the hen pen, which had seen much better days; wooden patches were nailed together to keep it from collapsing. Sure enough there were badger prints close by and one look at the pen told me a badger had tried to bite its way into one corner. There were teeth marks on the boards and large chunks of wood bitten-off, but the resulting hole wasn't big enough to let the badger in, so it had climbed up about 1 m to an old window covered only by rotten wire. There it had pushed its way in, leaving claw marks on the moss up the side of the hen pen. Inside were four dead hens. On the other side of the pen were more claw marks where the badger had obviously climbed its way out through the other window. "There, see," said the farmer. "He's eaten all our hens. We've lost more than a hundred." I looked round and could see only the four. Each had had its heart, liver and intestines eaten, and there were numerous badger hairs on these carcasses, but the heads were untouched – just what I would expect from a badger. So, where were all the others, I asked. The answer came quick enough. "He's being doing this for years. He must have killed over 100 in the past 40 years."

Well, I had to smile. Memories are long in the farming world. I decided to press him for more detail of this 40-year-old killer badger. "When does it happen?" I asked. "Mostly in the spring," he told me. Questioned further, he agreed it was usually after a cold snap. "Often when I've had to break the ice for the poultry." It was no comfort to him, but it did prove a point for me: this was an example of predation born out of desperation and made easy by dilapidation.

Another time, in south Shropshire, I was called out again by the RSPCA to an even more run-down small farm owned by an even older couple, both well into their eighties. The buildings were all but falling apart and only the cobwebs seemed to be keeping some of them together. A badger had killed a hen in its pen in the orchard. "I'm going to shoot this damn badger," the old man told me. "I've had enough." It was his wife who had called the RSPCA, but he was adamant. "Waste of time, you people are."

As it happened, I had brought an electric fence with me, as it would keep the badgers away and I told him, "Look, we can put this round your pen and that'll keep the badgers out." I explained it wouldn't cost him anything, as I was doing this as a goodwill gesture. "No", he said. "I'm not having that. That won't keep 'im out. I'm going to shoot it." "That's up to you," I told him. "But you'll be in trouble if you do. It's a protected animal, and you'll be breaking the law."

I had a look at the pen. Like the rest of the farm, it had seen better days. There were just two sticks propping the door closed and of course, no way would that prevent a badger getting in. I told him as much. Quick as a flash he came back, "I haven't shut the door to keep the damned foxes and badgers out. I've shut it to keep the hens in." There was nothing wrong in that bit of logic so far as he could see but, of course, it was the root of his problem. People with poultry have to accept that predators such as badgers, foxes, stoats, mink and polecats will follow their natural instincts and from time to time, will try to get to prey. So, they have to take reasonable care. But not my farmer friend: he wasn't going to accept advice. "I'm going to shoot that damn thing in the morning," he repeated. I looked at him. He was old, unsteady, shaking like a leaf. To keep the peace, I said, "Yes, I think that's the best thing you can do." So I left him, confident that he hadn't a cat in hell's chance of shooting straight enough. The badger would more likely die of laughing than of gunshot wounds.

Let me tell you about one more incident involving poultry. A badger got into a well-protected pen and killed some hens. I was too busy to go, but I talked to the owner on the phone. "Are you sure a badger is doing the damage and not a fox?" I asked him. "'Course I'm sure," he said, "it's still there, asleep in the nest box. I want it out." "Why don't you just leave it there," I offered hopefully. "It will go of its own accord tonight."

No, he was insistent he wanted it moved, so an RSPCA inspector and a police officer went instead. But, try as they might, they couldn't get the badger to move. Eventually they gave up, leaving the pen door ajar. Later, just as I had promised, the badger roused itself and ambled off into the night. Afterwards, I talked to the RSPCA inspector, "What was the problem? Why couldn't you get the badger out?" "We put the catcher noose round its neck" he said, "but the old beggar just dug its feet in against a wooden ledge and wouldn't budge. If we'd pulled any harder, we'd have pulled his head off." "You wouldn't have done that," I said, "He'd have given in before that. But, you know what you should have done: you shouldn't have pulled him forward. His front legs were working like a wedge. All you had to do was turn his head round and he would have reversed out. In a confined space, animals will always move backwards rather than forwards when frightened." As a bit of homespun advice, it had come too late, but I pass it on for anyone else faced with a similarly apparently intractable problem. That said, my recommendation is emphatically this: don't tangle with a trapped or injured badger, call in the experts – a powerful, frightened, injured badger is a hazard on four legs, an accident waiting to happen and you could be the accident! Leave well alone and fetch help.

While on the subject of badger casualties, I can't resist this observation: one of the odd things about people who suffer problems, aggravation or damage from badgers is that, all too often, they seem to think that those who try to protect badgers are the ones who should pay them compensation. Ask any badger group and they'll tell you it often happens. I well remember a telephone call one Sunday morning from a young man who was very annoyed. He said, "I understand you're something to do with badgers." "Yes, I am," I answered, "What's the problem?" He said, "My car has £800 of damage, thanks to one of your badgers. Who do I send my claim to?" "What happened?" I asked him. "He ran across the road in front of me and I hit him," replied the caller. "OK, then," I said, "Surely the normal procedure in road accidents is for the injured parties to exchange addresses and insurance details. I suggest you put your claim into the badger's insurance company. I take it you exchanged addresses?" With that, he put the phone down. Well, it was as daft an answer as the question he was asking! He thought that a wildlife organisation should pay for his car repairs – can you believe it?

The truth is, of course, that far too many badgers are killed unnecessarily on our roads. When I pick up badger casualties, I look on the road for the brake marks of the vehicle involved and all too often there are none. That upsets me and recently when a young badger cub was run over near my home, I placed a sign by it, which read: "If I had been an elephant, you would have stopped!" On another occasion, I put one up which said: "What if I had been a child?"

Carry on farming!

Ever since the Protection of Badgers Act 1992 came into force, the number of people seeking help with badger-related problems has increased and consultants and badger groups are often asked for advice on two particular types of problem: those involving badgers which have caused damage in gardens, and those where farmers are concerned about the spreading of badger setts.

Most farmers are prepared to put up with badgers in hedges and woodland; indeed, many actually enjoy having them on their land and are keen to protect them. The problems start when the badgers begin to dig shallow tunnels outwards into fields. It is then that holes begin to appear, say 10–20 m from the fence line, creating an apparent danger to livestock, though I have to say that in 50 years of farming none of our stock was injured that way and I think that such risks tend to be exaggerated. Traditionally, the advice given by conservationists (and for a while, many years ago, I went along with this) was to fence off the holes and leave that bit of land to the badgers. After all, for most farmers the odd corner or small piece of land is no great loss. But, my attitude has changed and I'll tell you why. The badgers inevitably start to go out further and further, and when this happens even the most sympathetic farmer begins to have second thoughts, as he ends up losing an unacceptable amount of land.

The tactics I now recommend are the result of practical experience. From the early 1940s right up until 1999, our own cattle grazed and browsed around badger setts, avoiding the holes which from time to time appeared some 5 to 10 m from the hedge. Occasionally, I put a piece of plywood over the hole, but eventually what I did was simply to pull grass away from the hole so as to make it very visible to the cattle. This seemed to work, and I didn't experience any problems. After a time, I didn't even do that, reasoning that the animals do have a sense of danger and are also observant, so I started to leave the holes alone and with the occasional exception of shallow tunnels near the surface that created the odd problem, this worked well. My advice, then, is do not fence off, but rather leave the sett as it is and let general wear and tear decide how the sett will develop. In my opinion, there is never going to be a serious danger of a dairy cow breaking a leg in a badger's hole and I have yet to hear a farmer tell me that their cattle have suffered any such injuries. Cows are much more likely to be injured slipping on the concrete in farmyards.

On arable land, the same sort of dilemma occurs when badger holes appear in land that is being prepared. Farmers who are sympathetic to the badgers pull the plough out and leave an unploughed area around the holes. But, inevitably, they face the same problem as the man who has left a bit of pasture fenced off and, as the holes spread further, the farmer loses more and more land. Eventually, he says, "That's it. I'm not putting up with this any more," so he ploughs the lot up. By contrast, you get the farmer who ploughs right to the edge every time, irrespective of whether holes have appeared. In fact, this used to happen in the field where I watched badgers for more than 30 years. The sett is still there and the badgers are still there. I have seen them pop up from other holes within hours of the plough going over the outlying holes and I think the message is clear enough. Only the very shallow tunnels in a sett are lost when ploughing occurs and having examined the interior of well over 150 setts as they have been dismantled, my view is that these shallow tunnels are soon replaced. Most, so far as I can judge, have just come about accidentally; the important part of the sett system is much deeper – usually over a metre or more down – and is not endangered by normal ploughing. It is no great loss to the badgers to lose a few new tunnels and the badgers themselves are in no danger. The sleeping chambers will almost certainly be well down, out of the way.

I know of one sett that spreads out on both sides of a hedge and the land on each side is owned by two different, but very efficient, farmers, both very economical with land. The boundary hedge is no more than 3.5 m wide and the distance from the furrows nearest the hedge on either side is no more that 6 m. There are two long-established badger entrances, one on each side of the hedge. Walking the land in July after the crops have been taken off, but before they have been re-ploughed ready for the next crop, I have found several new holes some 15–20 m out into the fields on each side. A few weeks later, the ploughs return, those holes in the field disappear and once more all that is visible of the sett are the two holes in the hedge. That has been going on to my knowledge for 30–40 years and the badgers are none the worse for it. They stay, and the farmers work their land and no remedial work is necessary.

Another example helps to make the point about what is the best solution. I was called out by a farmer who, for the past 6 or 7 years, had been advised to fence off land where holes had appeared and eventually had to take his fencing some 40 m out from the hedge, with the result that he had lost about half a hectare of grazing land. That was too much even for him and the time had come for a different approach, so he called me in. The badger sett was by now, of course, established over a very large area, and putting it right was a big job. We had to close the entire sett that was in the field under licence, leaving the badgers only what remained in the hedgerow and the main sett in the wood close by. This was very expensive for the farmer and very stressful to the badgers. Had he gone about his normal farming, spreading manure, cutting the grass for silage or hay, the tractor and trailer might occasionally have sunk into a shallow tunnel and caused indentations, but the badgers would have lived with it, the farmer would have been much happier and major remedial work would not have been needed. Deprived of the opportunity to expand outwards, the badgers would have developed the sett by going deeper or by pushing along the line of the hedge.

Just a quick word about horses. If anyone has horses in a field that contains a badger sett, and this is especially true if the horses are worked or galloped there, I suggest they get in touch with DEFRA. Horses are much more prone to break a leg if they stumble into a hole and then they would most likely have to be put down. DEFRA are knowledgeable, professional and unbiased about the risks and the remedies where horses are concerned and this type of problem, in any event, is an agricultural issue.

I have had lots of calls from farmers in my time, but even more from gardeners. I well remember one dry summer when I was inundated with calls from people troubled by badgers rooting through their lawns and flowerbeds. The first thing I tell callers as soon as they outline the problem is, "Stop watering your garden." "How do you know we are?" comes back the reply. I say, "Well, if you weren't watering, you wouldn't have a problem with badgers." The badgers' keen sense of smell tells them that there is dampness in the air and if the weather is dry, they know they are going to find food and good foraging where it is damp, so they make a beeline for that newly watered lawn or flowerbed. It is often claimed that, in an unusually dry summer, we lose a lot of badgers because they simply cannot find enough food to eat. I used to believe this until one very hot summer I had a little injured sow brought to me; she was only a fraction of the weight she ought to have been. Yet, other badgers brought in at the same time were normal weight, but this little sow had clearly not eaten for days. Then I noticed something

– she was the most heavily lice- and mite-infested badger I had ever come across. Internal or external parasitic infestations in a badger (she was wormed shortly after being rescued and found to be heavily infested) are much more debilitating in the summer and I suspect that badgers found dead or badly emaciated in hot, dry weather have died, not from lack of food, but from infestation by parasites.

A thin badger in July may be a sow that has just weaned a litter of cubs, as all multiple-birth mammals lose a large amount of weight at the end of lactation. In dry summers it is quite normal for badgers to compensate for the scarcity of worms and grubs by eating more grain and fruit, vegetables and berries. It is my experience that there is always more food available in the countryside than there is wildlife to eat it. But back to those gardening problems. If the badgers are busy in your vegetable patch, flower border or lawn in spring or autumn, then they are probably after a delicacy – either the larvae of cockchafers (maybug larvae) or crane-flies (leatherjackets). As for remedial measures, there isn't a great deal in these circumstances that you can do I am afraid. You can saturate your lawn with slug-repellent liquids but, if you do that, you are also going to kill off virtually everything and birds will certainly suffer. We must also remember that badgers have roamed the land for thousands of years before we invented gardens.

So, there are no easy solutions, not even for badger consultants. I well remember doing the labouring for my wife (a very keen gardener) to dig over and prepare a new plot of ground. She bought a lot of rare plants and she also used plants raised at home. She planted them all in one day and went down to the garden the next morning and let out a cry of despair. The plot was a shambles, everything had been torn out, plants were strewn everywhere, bent, twisted, ruined. Badger paw marks were also everywhere! That was really heart-breaking; my solution in this case was to put a generous helping of peanuts round the plot and to replant. The badgers took the nuts and fortunately, left the plants alone. You don't have to buy peanuts though – scraps of bread, leftover vegetables, almost anything edible will do (except meat, as it can spread disease). As long as their visit to your garden has been worth the effort, there is a good chance they will toddle off happily after feeding. Damage to lawns, in my view, is not quite so devastating, and it doesn't take long, if turf has been uprooted or turned over, to press the sods back into place and a lawn soon recovers. It is annoying, but the damage is reparable. That is the best advice I can give you. There is no absolute solution, but whatever you do, if you are keen on gardening, do not be tempted to build an artificial sett so that you have badgers all year round. That is sure to lead to frustration, bad tempers and regrets and turn your neighbour into an angry neighbour from hell.

Bovine tuberculosis (bTB)

Over 60 years ago a respected vet said to me: "To get a disease you have to create the conditions". In my lifetime of working with animals on a daily basis I have found this to be very true and it is especially relevant when discussing today's bovine TB problems.

I want to write simply from my perspective as a countryman and a farmer, so I'm not going to try and summarise all the scientific research into bovine TB. But if you would like to look into this, DEFRA's website has all the up-to-date research and information.

First let me explain briefly what TB is, then I would like to contribute my own down-to-earth views on this highly complex subject. In humans, tuberculosis (TB) is a serious slow-growing bacterial infection. It usually affects the lungs, causes serious breathing difficulties, general deterioration of health and can lead to death. Bovine tuberculosis (bTB) is a strain of TB that infects cows, badgers and other warm-blooded mammals. The bacteria that cause bTB (called *Mycobacterium bovis*), are very similar to

Mycobacterium tuberculosis, the organism that usually causes TB in humans. However, *M.bovis* can also cause tuberculosis in humans. Often a TB carrier can look and feel healthy and symptoms don't appear until the advanced stages

During my school days bTB was much more widespread than it is now. Parents lived in fear that their child would get this dreaded disease. It took more lives than any other contagious disease of that time and it killed slowly. Many suffered with it for years before eventually losing their battle. My brother contracted TB and spent 16 months in hospital. He was extremely poorly and our whole family were very worried. But we were lucky and he survived. Several of my school friends also contracted the disease. Some recovered, but others sadly died.

We were routinely tested at school for TB, and the treatment in those days placed great emphasis on fresh air and the beds in the TB hospitals were open to the elements. Most farm herds contained some TB, and one of the reasons why most farm children contracted TB was because they drank milk straight from the cow, without it being pasteurised. They would also play in the cattle sheds, which would have been contaminated with TB. I remember well that, before we had a milking machine, milking was carried out by hand and all family members would be involved, and very young children would be taken into the cowshed and put into an empty stall that was used as a playpen, thus exposing them to TB.

During the 1930s the Ministry of Agriculture Fisheries and Food (MAFF) started to test cattle for bTB. Montgomeryshire (now part of Powys) became the first county in the United Kingdom to be free from bTB. By the early 1960s, bTB had been virtually eradicated from the British Isles. If this could be achieved back then surely it can be now. Our own herd was free from bTB by 1950, and remained so until its dispersal in 2008. Now I'd like to tell you how it was done.

Testing & culling cattle. All cattle were tested on an annual basis for bTB and those that reacted positively were culled.

This may sound simple (and it is!) but there's been a huge decline in testing over the last 20–30 years and some cattle live their entire lives without ever being tested for bTB.

Movement of cattle. Moving cattle from farm to farm was strictly controlled from the 1930s to the early 1960s which meant bTB couldn't easily be transmitted to cattle on other farms. No cattle were allowed to be moved to other farms until it was proven they were clear of bTB. As a county became free of bTB no cattle from an infected county could be moved into that county. These strict guidelines are not applied today. Also the boundaries were double fenced so the cattle from different herds couldn't touch one other.

Cleanliness. All farm buildings needed to be thoroughly cleansed and a MAFF inspector would come out to inspect these. We scrubbed all cattle housing by hand with caustic soda. I remember having to help and crying with pain as the caustic soda water took the skin from my fingers - we had no rubber gloves in those days!

Ventilation. Fresh air is vital to keep TB at bay. It is my experience that larger buildings with more stock lead to the creation of pockets of poor quality air which I feel increases the risk of disease being transmitted. In my day, glass was removed from livestock buildings to ensure good air movement. Why are such measures not practised today?

Bloodlines & breeds. Most of our infected cattle were from particular family bloodlines that were more susceptible to bTB. During the eradication process we lost two bloodlines on our farm, one of which was our favourite due to their docile temperament and good looks. By the 1960s, the bloodlines that were most susceptible to TB had been culled. This left healthy herds of cattle that had a resistance to TB, such as British Friesians, British Shorthorns, Ayrshires and Herefords.

Unfortunately modern farming methods have seen a huge decline in these breeds of cattle. They have been replaced by continental breeds which don't appear to have the same level of resistance to bTB.

Herd size. When bovine TB was virtually wiped out by the early 1960s it was from much smaller herds. Today's herds are much larger and that makes it more difficult to find and remove all infected animals, so it's more likely that infection in a herd will continue to spread, especially as the 'live' test is flawed. I also feel the winter housing used nowadays is unsuitable. Packing more and more animals into larger buildings for months on end creates extra stress on cattle and stress can weaken immunity to diseases.

Farming has changed out of all recognition since those postwar labour-intensive days, but the lessons learnt from that epidemic are just as relevant today. So, too, is that bit of veterinary wisdom: to create a disease you have to create the conditions. The way cattle are housed, fed, and subjected to stress, and of course the circumstances and the frequency in which they come into contact with other cattle are all potential underlying causes of bTB.

Badgers and bTB

Let's turn now to the subject of badgers and bTB. Knowing and understanding the feeding habits of both badgers and cattle, I do not understand how badgers can be considered responsible for passing bTB to cattle. To me it seems more likely that it is the other way round. When cattle graze, they leave large amounts of saliva on the grass, and if the cattle are infected the saliva will be, too. When badgers forage where cattle have been grazing it is almost inevitable that they will ingest some of this saliva and with it the infection. Badgers often search for grubs under cattle dung, so that's another obvious way they could pick up infection.

Bear in mind that bTB is primarily a lung disease and it's most likely to be passed on when tiny droplets of infection are expelled - normally on the breath or by coughing. Could badgers infect cattle that way? I think we can rule out that possibility. It's said that urine of infected badgers contains TB, which could in turn infect cattle, but the risk of that happening is surely very slight. Cattle far outnumber badgers on a farm. On the same area of land you might have, for example, 100 or more cattle and only two or three badgers. To put it in perspective it is estimated that there are around 300,000 badgers in the United Kingdom and approximately ten million cattle.

I think it's also worth noting that sheep spend a great deal of time lying on, or close to, badger setts. They also spend more time in close contact with badgers than cattle do and yet there doesn't appear to be a problem with TB in sheep.

There does not seem to be a significant TB problem with foxes, either. Again, they differ from cattle and badgers by spending a great deal of time in the open air. They spend most of their time above ground, and when underground are not very far from an entrance. Deer are bovines, yet do not appear to have a high incidence of TB. Could this be because they are not subjected to the high levels of stress experienced by cattle?

Bovine TB is constantly in the news, as are farmers complaining how many cattle they lose to bTB. But in reality the number of cattle in England and Wales being infected with bTB has dropped. In 2008, 40,000 were slaughtered due to bTB. This decreased to 36,000 in 2009[1]. Compare this to cattle slaughtered annually for other ailments. In 2003 for example:

125,000 due to infertility[2]

90,000 due to mastitis[2]

31,000 due to lameness[2].

As you can see, farmers are losing far more of their cattle to other ailments which cause a greater financial loss than bTB.

My main point is that the badger's lifestyle has not altered for many thousands of years. Disease or no disease, they have survived. In the cattle industry, the increase of bTB has happened in parallel with the changes in management of dairy and beef cattle and if I were in farming today that's where I would want the research concentrated.

---

[1]www.farmersguardian.com/home/hot-topics/bovine-tb-(btb)/32043.article

[2] Sibley, R. (2003). Rethink health strategies. Farmers Weekly. February 28th, 2003.

Chapter Eight

Badger-watching

Sett-watching tips

Finally, I should like to offer some advice about badger-watching. Unless you are extremely dedicated or very lucky, most of your best sightings will be close to setts as the light fades. Here are a few basic rules. Remember always to ask permission from the farmer or landowner. It is important first to visit the sett in daylight to get the lie of the land and to select up to four possible vantage points. Then, whatever direction the wind is blowing, you will be able to sit downwind so the badgers do not pick up your scent. When possible, sit facing east, as that way you will have the light behind you.

When approaching the sett, take your time and savour the evening air. Listen for the sounds of the various birds and mammals – the sweet song of the wren, the alarm calls of a blackbird, the distant mooing of cattle, small mammals rustling in the leaves. Also, take in the different smells – the perfume of plants and the scent of various animals, which are much stronger at night and in the early morning. You will find yourself ducking your head as bats swoop very low. Stop and stand still for a few moments and slowly look around, as you may well have disturbed a fox as you were walking. It will hear you coming and move a short distance out of the way, then watch you pass and you won't be aware of it. When you are out of its danger zone, it will return to the exact spot where you disturbed it and watch you disappear into the distance.

Only experience will teach you how close to the sett you will be able to watch without disturbing the badgers. This will vary from sett to sett and be determined by factors such as the amount of natural cover, the way you behave and whether the badgers are used to the smells and sounds of humans. With some badgers you can sit as close as 5 m, with others you may be 25 m away and they will still be shy and nervous, but a good rule-of-thumb is to be at least 10 m from the sett. Always try to sit under the cover of a hedge, bush or tree. Take an old stool or chair and a comfortable cushion. Polystyrene cushions in winter are wonderfully warm and a piece of polystyrene under your feet is great in frost and snow. You need to be comfortable, so do not wear garments that are so tight they make you fidget. Wool is warm and it doesn't rustle; remember, silence really is golden. Avoid bright colours and for obvious reasons do not drink too much liquid before you set off! Let someone know where you are going and how long you expect to be. Generally, two people at a time is the maximum for a sett in the wild where the badgers rarely encounter human disturbance, but you will find the most rewarding badger-watching will occur when you are on your own.

You may get the chance to watch badgers where they are fed regularly, possibly from a purpose-built hide. There are numerous places around the country that provide badger-watching facilities – these are advertised in most wildlife magazines or websites. That's fine – it can be fun and it's a great way of seeing them close up, watching how they feed, how they search with their noses for food. But, if you want to see true, natural badger behaviour, then find yourself a quiet out-of-the-way sett that is rarely disturbed by humans. Badgers that are regularly fed and are used to human scent will make a beeline for the food as soon as they emerge from their sett. You get instant action, but what you miss is the natural caution of the emergence and the anticipation and excitement that goes with it. Alone at night by a sett, I have often seen in the gathering gloom some sort of movement, some little speck of light that would appear and disappear. Was it a badger or a figment of my imagination? Did it get my scent? The tension is part of the pleasure. At one sett, I didn't solve that dilemma for certain until one night I decided to take a flashlight photo of what I thought was a figment of my imagination. When I got the prints back, the answer was there. Yes, what I'd seen was a badger's snout cautiously scenting the air. I concluded that, for some years, the same had been happening at that sett without me knowing.

On a mild night, you can hear them moving underground before they emerge. Expect the odd whicker or grunt and quite often a whiff of their scent, a musky smell a bit like rotting apples. I've known this to go on for over an hour. Then they emerge, sniff the air and move off straight away, having done most of their stretching and scratching underground – it is common to find chambers in the sett large enough for vigorous activity. The first badger out will invariably pause, listen and turn its head, and you can tell whether it is taking in scent or listening. Commonly, badgers will groom each other or scent mark, the dominant boar and sow scent marking the subordinate ones. Often, scent marking seems to be a free-for-all as they mark areas around the sett, stones or the base of trees. I and other badger watchers have had our feet scent marked because we did not smell like a badger. This sort of preparation can go on for several minutes.

In one of the wilder areas where I used to watch badgers, they would often leave very quickly and the badger-watching didn't last long. So one day I decided to follow one of them. It went under the fence, so I tried to climb over. I managed that one, but at the next one I tore my trousers and I fell head over heels. He heard me, of course and trundled off back to the sett. A week later, I tried again, having first had a look at the fence in daylight to find a better way over. I was able to follow him through the fence, across a field of barley, over a road, across another field of barley, over another road, across a stream and into an old pasture where he began to turn over cowpats. I thought, this is great, time to get a photo and I started clicking away with my camera to try to get a record of him doing just that, especially the sort of pancake-tossing effect badgers go in for when they're absorbed with this source of food. Unfortunately, just one barely usable photo came out of that session, as my hands were too shaky. However, it does show that, if you are determined enough, then tracking badgers is possible – provided, of course, the wind is in the right direction. One whiff of your scent and that's it: they're off.

Sit quietly!

Wherever you choose to watch badgers, remember that they have exceptionally acute hearing. They pick up sounds we cannot hear, at times reacting to just the slightest rustle, the merest whisper and yet they will put up with a barrage of noise – anything from exploding bombs and jet planes to trains and roaring traffic. What is important to them and therefore what is important to the badger-watcher, is whether the noise is something they are used to or whether it is different or new and therefore in some way a possible threat. I watch badgers at a sett close to army training grounds and often the still of a summer's evening is broken by the sound of rapid rifle fire, the blast of hand grenades or the heavier thump of bombs exploding. The badgers are totally unconcerned; they simply carry on grooming and foraging. But if I pull a handkerchief ever so gently from my pocket or inadvertently snap a twig underfoot, then they vanish underground; the pattern of sound has been broken and their instinct is to bolt for cover. Badgers will sometimes choose to build their setts underneath roads, near railway tracks and even airfields, where the noise and the rumbling of the traffic, planes, helicopters and trains must be horrendous, yet the badgers seem unconcerned. Badgers living on the edge of quarries tolerate blasting within metres of their sett, choosing to put up with all that noise, heat and vibration. Compare that with badgers in a typically quiet sett, where the merest whisper can make them disappear. Perhaps the pitch of the sound has something to do with it, and maybe the whisper carries with it a sense of anxiety or alarm. More likely again, it is simply that the sound is different, one they haven't become accustomed to.

The weather

Let us now think a bit more about the weather, and what it means to the badger-watcher. Let us say, for example, you are getting dressed to go out for an evening's badger-watching, you look out of the window and it starts to rain. Do you go? My advice is yes, it could well prove to be one of your best nights. I've found out over the years that, in warm, steady rain, badgers are less cautious. There's something about the wet they seem to enjoy…an abundance of earthworms! High winds, provided the weather is warm and mild, will have no detrimental effect. But a cold east wind? Stay indoors and put your feet up. As Corporal Jones of *Dad's Army* might put it: 'badgers don't like it up 'em'. They abhor cold east winds and are not very partial to cold northerly winds either. I have had very few sightings in easterlies. Over a six-year period, I watched badgers on no fewer than five nights a week, even Christmas night. I watched them through all seasons and in all weather conditions and during all that time the cold easterly winds were the worst. What about snow? I've seen them out in icy conditions with snow covering the ground and the temperature seven degrees below freezing. In cold weather I always took a temperature gauge with me. One very cold night, I had to wait only 20 minutes before three came out. I took a whole roll of film that night, but when it was developed there was only one that was of quality. Some time later, I discussed this with a photographer and he said what a pity I hadn't put three pairs of socks on the camera as the cold had affected the very fine workings of it. The night before, it was even colder – 9 degrees below freezing and I dressed accordingly: three pairs of socks, three pairs of trousers (including over trousers) and three pullovers underneath my weatherproofed coat! I stayed out for just an hour and saw nothing during that short stay, but returning next morning found plenty of prints in the snow. They had certainly ventured out.

Feeding badgers

What about feeding badgers to encourage them to stay while you watch? I have no qualms about badgers being fed every night if that's what people want to do. But they should not be overfed – titbits rather than major meals. Just remember, you are influencing their behaviour and it is not natural. One sett I used to sit by always produced activity very early in the evening, often with an hour or more of light left. I could observe them, go back home, have a snack, visit another sett and the badgers there would still not have been out. The wilder and more remote the location, the more likely I feel that they will come out early. They are rarely if ever disturbed, so they have no need to stay underground simply to be safe. At setts more often visited or close to people or paths, emergence can often be quite late. In some setts the badgers rarely, if ever, emerge during daylight. These are broad behaviour patterns that occur quite commonly. But I come to something I've often said: badgers, like people, are individuals. Some are bold, some are nervous, some skittish. You have to take all these variables into account. That is one of the pleasures of watching wildlife – you can never be sure what the animals will do.

Seeing in the dark

What about torches? Some people don't like the dark and prefer to find their way using a torch. I am the opposite and I would rather do without one if I can. It is amazing how your eyes become used to almost no light in a short time. I have used torches with red filters, particularly if I have been trying to fix a spot to focus on for photographic purposes. But one night I took an ordinary torch instead and to my surprise the badgers seemed unperturbed by the light. I used to worry about flashlight photography, but it never seems to bother them. They are more concerned about the click the lens makes, simply because it is an unfamiliar noise. I have also found that good-quality binoculars will 'extend' natural daylight by almost an hour on certain moonlit nights.

When night vision image intensifiers first became available, I bought one and it really transformed my night-time viewing. Viewed through the night vision equipment, badgers are easy to see. The infra-red beam appears to light up the night like a beacon, but in fact is totally invisible to badgers, so they continue to act entirely normally. The first night vision scope I used was an ex-Russian military scope. That was in the early 1990s and the equipment available today is much more advanced. When buying night vision equipment, there are no hard and fast rules and you don't have to spend vast amounts to get the results. There are some very good pieces of night vision equipment available, at prices ranging from under £200 for entry-level equipment to several thousands of pounds for the latest military surveillance specification equipment.

Hides

Hides can be useful. One of the best is a car or vehicle of some kind and I often watched from a tractor. I've never ceased to be surprised, when I've been working land from a tractor, how close you can get to wildlife; they simply hop a few metres away and seem unperturbed by the noise or the movement. I spent endless hours half hidden in a battered old wreck of a car that lay discarded not yards from my favourite sett and this proved an ideal hide. Occasionally, foxes, rabbits and polecats appeared, owls called from the trees nearby and bats swooped in from all directions, but it was the badgers that drew me to that spot so often. They gave me immense pleasure and I hope that this book encourages you to find a favourite sett or two and to while away the hours in the company of these wonderful wild animals.

Chapter Nine

Final thoughts

So much for my experiences, my reflections on a lifetime spent watching, learning about and caring for badgers. If you are already 'into badgers' I hope the book has been an enjoyable read and that in some small way it has helped you understand more about these fascinating animals.

If you haven't yet started watching, if you haven't yet found a sett you can sit by, please do make the effort. Look for the clues, especially those spoil heaps and broad badger paths. It really is worthwhile. Believe me, the memory of your first sighting, the moment when just yards away a black and white head emerges cautiously from a sett to scent the air, is one you will never forget.

Memorable, too, are the days when the cubs appear for the first time. Inquisitive, full of mischief, bold and yet fearful, quick to explore and just as quick to run back to the safety of their sett, they are a joy.

So sit quietly, be patient, relax. Let nature come to you. Enjoy the warmth of a late spring or early summer's evening. As the light slips away, the woods and the hedges will be full of song. The more often you watch, the more you will see: foxes as they start to hunt for food; deer emerging from the safety of the woods where they've rested all day; woodmice darting here and there, grabbing a peanut meant for the badgers and disappearing at speed down the nearest bolthole. Owls, bats, late evening rooks returning to roosts, strutting cock pheasants, and in my part of the world, even the occasional polecat.

Enjoy every moment. One thing is certain, however much you know, however often you've sat and waited and watched, badgers will sooner or later provide you with another special memory to tuck away, to relive, to smile about. There's always something new to learn and there's no doubt that in the years ahead scientific research will continue to shape our understanding of this ancient, secretive creature.

For my part, I still worry about the future and what lies ahead. Badgers have survived for millions of years, but at the hands of man they have suffered and still suffer horrific abuse and persecution. Legislation has helped to reduce crimes against badgers and their setts, but of course unless laws are actively enforced much of their deterrent value is lost.

Until every police force has at least one dedicated wildlife crime specialist, the badger will remain vulnerable, constantly under threat, reliant on the vigilance of the public and especially conservation volunteers. Badger groups were formed to protect badgers and they continue to work hard safeguarding setts, monitoring suspicious activities, and working hand in hand with agencies like Natural England, the police and the RSPCA – whose special operations unit has brought many a badger persecutor to justice.

Collectively the groups formed the National Federation of Badger Groups which has now become the Badger Trust. Although I work independently of all wildlife organisations I very much respect the work they do. Faced most recently with the prospect of state-funded mass slaughter of thousands of badgers, Badger Trust has argued its case so successfully that a proposed cull by the Welsh Assembly has been declared unlawful by the Court of Appeal and for the time being at least that cull, part of the Assembly's plans to tackle bovine TB, has been stopped.

However, Badger Trust and most other conservation bodies concerned about the fate of badgers, regard this merely as a battle won. The struggle to protect badgers both from persecution by individuals and by the state, goes on. Now more than ever, badgers need friends, people who will look out for them, fight their corner. So if you can, please do your bit to save our wildlife for future generations.

But enough of the sermonising. I'd like to end with just one more memory. It's fresh in my mind because it happened as I was working on the closing chapters of this book. It was 16th February 2010, bitterly cold and the ground was covered in two inches of snow. I was at my home in the Shropshire countryside, working late but keeping watch for the first sign of the sow that regularly visits for the peanuts we leave out. This time she wasn't alone. This time she had not one, not two, not three, but four cubs with her. They were sturdy cubs, by my reckoning at least 10 weeks old, which means they had been born about mid December.

That's much earlier than usual. The broad rule of thumb, you will recall, is for cubs to appear above ground for the first time in early to mid April. But that's nature, full of surprises, always breaking the rules. And what a joy that is. It is only the second time in my lifetime that I have seen cubs as early in the year as this.

This farmer's boy is old in the tooth now. The back creaks and the joints complain. But the wonderful memories live on and I feel very privileged that wild animals have allowed me to share part of their lives with them. I hope many of you will feel that way too.

Useful contacts

Campaign to Protect Rural England (CPRE)
www.cpre.org.uk
T: 020 7981 2800

Countryside Council for Wales (CCW)
The Government's statutory advisor on sustaining natural beauty, wildlife and the opportunity for outdoor enjoyment in Wales and its inshore waters.
www.ccw.gov.uk
T: 0845 1306 229

Department for Environment, Food and Rural Affairs (DEFRA)
www.defra.gov.uk
T: **08459 33 55 77**

Natural England
An independent public body whose purpose is to protect and improve England's natural environment and encourage people to enjoy and get involved in their surroundings.
www.naturalengland.org.uk
T: **0845 600 3078**

Scottish Natural Heritage (SNH)
SNH is funded by the Scottish Government, with the purpose of: promoting care for and improvement of the natural heritage; helping people enjoy it responsibly; enabling greater understanding and awareness of it; promoting its sustainable use, now and for future generations.
www.snh.gov.uk
T (Head Office): 01463 725000

The Badger Trust
Promotes the conservation and welfare of badgers and the protection of their setts and habitats for the public benefit.
www.badger.org.uk
T: 08458 287878

The Mammal Society
Works to protect British mammals, halt the decline of threatened species, and advise on all issues affecting British Mammals.
www.mammal.org.uk
T: 02380 237874

The Wildlife Trusts
A voluntary organisation dedicated to conserving the full range of the UK's habitats and species.
www.wildlifetrusts.org
T: 01636 677711

Royal Society for the Prevention of Cruelty to Animals (RSPCA)
The leading animal welfare charity.
www.rspca.org.uk
T (advice): 0300 1234 555 / (cruelty): 0300 1234 999

G E Pearce Badger Consultants
Expert, independent and unbiased service for all issues relating to badgers and their habitat.
www.badgerconsultants.co.uk
T: 01939 260 600

Pearce Environment Ltd
An ecological consultancy providing expert ecological advice.
www.pearce-environment.co.uk
T: 01743 741 421

Wildlifeshop.co.uk
For all your wildlife products, from bird boxes to night vision equipment.
www.wildlifeshop.co.uk
T: 01743 741 421

Index

About Pelagic Publishing

We publish books for scientists, conservationists, ecologists, wildlife enthusiasts – anyone with a passion for understanding and exploring the natural world. Working closely with authors and organisations we are publishing books that:

- *Deliver cutting-edge knowledge, published rapidly in traditional and eBook formats.*
- *Promote best-practice in research techniques and management methods.*
- *Encourage and assist practical wildlife investigation and field exploration.*
- *Bridge the gap between scientific theory and practical implementation.*
- *Share wildlife experiences.*
- *Promote taxonomic and identification skills.*
- *Highlight the use of technology in science and wildlife exploration.*

To stay up-to-date with news and offers visit:

www.pelagicpublishing.com

A discount on your next book

Save 15% on your next order from Pelagic Publishing by quoting 'BNH15' when you order from our website (www.pelagicpublishing.com). Terms and conditions apply: this coupon can only be used once per customer and cannot be used in conjunction with any other offer.